Water Budgets and Rate Structures:

Innovative Management Tools

Awwa Research Foundation

About the Awwa Research Foundation

The Awwa Research Foundation (AwwaRF) is a member-supported, international, nonprofit organization that sponsors research to enable water utilities, public health agencies, and other professionals to provide safe and affordable drinking water to consumers.

The Foundation's mission is to advance the science of water to improve the quality of life. To achieve this mission, the Foundation sponsors studies on all aspects of drinking water, including supply and resources, treatment, monitoring and analysis, distribution, management, and health effects. Funding for research is provided primarily by subscription payments from approximately 1,000 utilities, consulting firms, and manufacturers in North America and abroad. Additional funding comes from collaborative partnerships with other national and international organizations, allowing for resources to be leveraged, expertise to be shared, and broad-based knowledge to be developed and disseminated. Government funding serves as a third source of research dollars.

From its headquarters in Denver, Colorado, the Foundation's staff directs and supports the efforts of more than 800 volunteers who serve on the board of trustees and various committees. These volunteers represent many facets of the water industry, and contribute their expertise to select and monitor research studies that benefit the entire drinking water community.

The results of research are disseminated through a number of channels, including reports, the Web site, conferences, and periodicals.

For subscribers, the Foundation serves as a cooperative program in which water suppliers unite to pool their resources. By applying Foundation research findings, these water suppliers can save substantial costs and stay on the leading edge of drinking water science and technology. Since its inception, AwwaRF has supplied the water community with more than $300 million in applied research.

More information about the Foundation and how to become a subscriber is available on the Web at **www.awwarf.org.**

Water Budgets and Rate Structures:

Innovative Management Tools

Prepared by:
Peter Mayer and **William DeOreo**
Aquacraft, Inc., 2709 Pine St., Boulder, CO 80302

Thomas Chesnutt and **David Pekelney**
A&N Technical Services, 839 Second St, Suite 35, Encinitas, CA 92024-4452
and

Lyle Summers
Utah Division of Water Resources
Water Appraisal Service Company, 3961 West 8010 South, West Jordan, UT 84088

Jointly sponsored by:
Awwa Research Foundation
6666 West Quincy Avenue, Denver, CO 80235-3098
and

City of San Diego Water Department
600 B Street, San Diego, CA 92101-4589

Additional funding provided by:
Pinellas County Water System
Jordan Valley Water Conservancy
City of Scottsdale
East Bay Municipal Utility District
City of San Juan Capistranoa
California Urban Water Conservation Council
U.S. Environmental Protection Agency

Published by:

Awwa Research Foundation

Distributed by:

American Water Works Association

IWA Publishing

DISCLAIMER

This study was jointly funded by the Awwa Research Foundation (AwwaRF) and the City of San Diego Water Department (San Diego). AwwaRF and San Diego assume no responsibility for the content of the research study reported in this publication or for the opinions or statements of fact expressed in the report. The mention of trade names for commercial products does not represent or imply the approval or endorsement of AwwaRF or San Diego. This report is presented solely for informational purposes.

CONTENTS

LIST OF TABLES

LIST OF FIGURES

FOREWORD

The Awwa Research Foundation is a nonprofit corporation that is dedicated to the implementation of a research effort to help utilities respond to regulatory requirements and traditional high-priority concerns of the industry. The research agenda is developed through a process of consultation with subscribers and drinking water professionals. Under the umbrella of a Strategic Research Plan, the Research Advisory Council prioritizes the suggested projects based upon current and future needs, applicability, and past work; the recommendations are forwarded to the Board of Trustees for final selection. The Foundation also sponsors research projects through the unsolicited proposal process; the Collaborative Research, Research Applications, and Tailored Collaboration Programs; and various joint research efforts with organizations such as the U.S. Environmental Protection Agency, the U.S. Bureau of Reclamation, and the Association of California Water Agencies.

This publication is a result of one of these sponsored studies, and it is hoped that its findings will be applied in communities throughout the world. The following report serves not only as a means of communicating the results of the water industry's centralized research program, but also as a tool to enlist the further support of the nonmember utilities and individuals.

Projects are managed closely from their inception to the final report by the Foundation's staff and large cadre of volunteers who willingly contribute their time and expertise. The foundation serves a planning and management function and awards contracts to other institutions such as water utilities, universities, and engineering firms. The funding for this research effort comes primarily from the Subscription Program, through which water utilities subscribe to the research program and make an annual payment proportionate to the volume of water they deliver and consultants and manufacturers subscribe based on their annual billings. The program offers a cost-effective and fair method for funding research in the public interest.

A broad spectrum of water supply issues is addressed by the Foundation's research agenda: resources, treatment and operations, distribution and storage, water quality and analysis, toxicology, economics, and management. The ultimate purpose of the coordinated effort is to assist water suppliers to provide the highest possible quality of water economically and reliably. The true benefits are realized when the results are implemented at the utility level. The Foundation's trustees are pleased to offer this publication as a contribution toward that end.

David E. Rager
Chair, Board of Trustees
Awwa Research Foundation

Robert C. Renner, P.E.
Executive Director
Awwa Research Foundation

ACKNOWLEDGMENTS

This research project was made possible by the generosity, interest, and perseverance of a great number of individuals who supported the effort from its genesis to completion. The research team wishes to offer our heartfelt thanks to everyone.

This project was made possible by AwwaRF and the following organizations that provided funding and in-kind support: United States Environmental Protection Agency, City of San Diego Water, City of Scottsdale Water Resources Department, Pinellas County Utilities, East Bay Municipal Utility District, California Urban Water Conservation Council, Jordan Valley Water Conservancy District, and City of San Juan Capistrano.

This project was aided tremendously by the in-kind contributions of the following organizations: Las Vegas Valley Water District, City of Portland, Otay Water District, Department of Water and Power the City of Los Angeles, Irvine Ranch Water District, Davis California Public Works Department, City of West Jordan, Centennial Water and Sanitation District, City of Boulder, City of Santa Barbara, Town of Cary, City of Aurora, and Marco Island.

The project team wishes to express our sincere thanks to the following individuals and groups for their assistance and contributions to this project: John Flowers, Marsi Steirer, Roy Martinez, Mary Ann Dickinson, Leslie Martien, Matt Hayden, Dana Holt, Tom Ash, Francie Kennedy, Cynthia Russell, Jean McGonigle, David Mansfield, John Ralston, Chris Staubus, Pick Talley, Wayne West, Barbara Kuhl, Richard Harris, Dick Bennett, Richard Bay, David Ovard, Doug Bennett, Kent Sovocool, Richard Wimmer, William Granger, Jan Tubiolo, Rick McLoud, Kevin Reidy, G. Hossein Parandvash, Mort Anoushinavani, Fiona Sanchez, Dave Ferguson, Paul Jones, Richard West, Robert Weir, Jaques DeBra, Jennifer Platt, Robert Fisher, Gary Luebbers, Richard Calvi, Paul Lander, Ned Williams, Carol Ellinghouse, Bronwyn Weigant, Jim Knopf, Jan Beecher.

The members of the project advisory committee (PAC) provided excellent input into this project at all stages of the effort. Our sincere thanks go to Eric Rothstein or CH2M Hill, David Bracciano of Tampa Bay Water, and Scott Rubin for all of the time, effort, and thought that they put into this project. Finally, we would like to thank our AwwaRF Project Manager India Williams for her assistance and guidance.

EXECUTIVE SUMMARY

"Water budgets"—volumetric allotments of water to customers based on customer-specific characteristics and conservative resource standards—are an innovative means of improving water use efficiency. Once thought to be impractical because of technological constraints, water budgets linked with an increasing block rate structure have been implemented successfully in more than 20 utilities. As utilities develop advanced customer information systems and geographical information systems these rate structures are expected to be applied more broadly. Water budget rate structures are attractive to water agencies searching for stable revenue generation, improved customer acceptance, increased water use efficiency, augmented affordability of nondiscretionary customer water consumption, and improved drought response. A growing number of utility managers are finding that water budgets offer potential benefits to water utilities and their customers in coping with increasing water scarcity and rising costs.

Water budget-based rate structures are probably not for everyone. This rate structure form requires more customer level data than traditional rate structures and may be more expensive and labor intensive to put in place. Utilities without a pressing need to encourage water conservation are not good candidates for water budgets because implementation costs might exceed any non-conservation benefits. For utilities that wish to supplement water conservation efforts, and send a fair and effective price signal about usage to their customers, water budgets appear to be an excellent option to consider.

This research project examined water budgets and their potential value to North American water utilities and the varying applications of the water budget concept that have been adapted to different conditions. Key issues identified include: different practical approaches to water budget rate structures, the benefits and challenges of these approaches, the potential uses of water budgets during drought, and important steps in the water budget implementation process. The case studies presented in this report offer first hand information on the design, implementation, and management of water budget rate structures as well as some cautionary examples of water budget programs that were rescinded or that never got off the ground.

Although water budgets have been implemented by water providers in North America for nearly 20 years, until recently they had only been used by a handful of utilities. As populations increase and climate uncertainties place heightened demand and stresses on water systems, more utilities are seeking new tools for water conservation and drought response. Water budget-based rate structures are one of the most effective tools available to water providers to provide a meaningful price signal, reduce water waste, increase efficiency, and manage drought response in a sensible and equitable manner.

Definition of a Water Budget

A water budget is the quantity of water that would be required by an efficient level of water use. Exactly what constitutes an "efficient level" of use (particularly as it relates to outdoor irrigation) is often a community decision and must be based upon public expectations as well as water supply realities. Water providers tend to define water budgets in more specific terms so that their customers can easily understand how the budget (or allocation as it is often called) applies. For example, the City of Boulder, Colorado defines the concept for customers in the following way: "A water budget is the amount of water you are expected to need for a specific month. Each customer's water budget will be different based on their water needs. Water budgets may vary monthly based on seasonal outdoor watering needs."

Landscape water budgets are the most common form of a water budget. A landscape water budget is typically a volume of water that is calculated from two parameters: the landscape size (usually in square feet) and the water requirement of plants in that geographic area which is usually represented by the evapotranspiration (ET) rate.

Indoor water budgets are typically a fixed volume provided each month based on the historic use of the customer class or a calculated volume based on the anticipated needs of the customer. Indoor water budgets may vary based on the number of people in a household, the type of manufacturing processes used in an industry, the size of a business, or other factors.

Water Budget-Based Rates

Water budget-based water rates—also known as individualized, goal-based, and customer specific rates—are block rates where the block is defined by using one or more customer characteristics. Water budget-based rate structures can be thought of as an increasing block rate structure where the block definition is different for each customer based on an efficient level of water use for that customer.

Water Budgets Can Meet Cost of Service Requirements

According to the AWWA M1 Manual, "the basic premise in establishing adequate rate schedules that are equitable to different customers is that rates should reflect the cost of providing water service" (AWWA 2000). After revenue requirements have been established, costs are allocated among different types of water users, and then rates are designed to reflect the cost of providing water service. Although the concept of cost of service is central to public utility rate setting, most rates reflect historical pricing regimes and local political imperatives. Legal precedents require that rates be "fair and reasonable." Public utility practice often uses a cost-of-service study to assess the reasonableness of rates.

A water budget-based rate structure is inherently no more or no less cost-based than a similar increasing block rate structure without a water budget component. There is little dispute that increasing block rate structures can be properly designed to meet cost-of-service requirements. The same is true for a water budget-based rate structure. The research team could find no report or publication that identified cases where water budget rate structures failed to meet cost of service criteria. Given that more than 20 utilities are actively implementing some form of these rate structures, it appears that properly designed water budget-based rate structures can meet these important criteria.

Water Budgets are Being Used Across the USA

During the 1990s, water budget rate structures were only implemented by a few water agencies in California, most notably in Capistrano Valley, IRWD, and the Otay Water District. By 2007, water budget rate structures have diversified and expanded throughout California and across the United States to Utah, Nevada, Colorado, North Carolina, and Florida. The researchers were able to identify 25 water providers that are using some form of a water budget for billing and informational purposes.

While still primarily a California and western states phenomena, water budgets have found support in a broad range of agencies from small systems like Marco Island, Florida to the Los Angeles Department of Water and Power (LADWP). Water budgets are no longer an exclusive "boutique" rate structure for select technologically savvy providers. These rate structures have been adopted by a broad range of providers, each adopting their own version of the water budget concept.

Water Budget-Based Rate Structures, Like Utilities, are Unique

Utilities are unique and so are their rate structures be they water budget-based or not. The formulation of water budgets and implementation within the utility context was unique for every example the researchers identified. Only three of the identified water budget utilities developed a water budget for all customers in the service area. Most of the water budget utilities opted to use water budgets for only certain customer classes (such as single-family residential or dedicated irrigation) and some only use water budgets for informational purposes in which the budget is not linked to the rate structure. Single-family residential and dedicated irrigation accounts were the most common customer categories to be included in a water budget program.

Landscape Water Budgets Are Most Frequently Implemented

Landscape water budgets were the original basis for the formulation of these rate structures and are the foundation of the majority of implementations examined in this report. The idea of calculating the volume of water required to adequately irrigate a plot of land is hardly a new concept. Farmers and agronomists have been developing water budgets for agriculture for many years using evapotranspiration measurements to determine the required application depth to maximize plant growth. In the urban context, water budgets have been used for years to determine irrigation efficiency for parks and large irrigated parcels. They have been used at least since 1989 in conjunction with inclining block rate structures to provide a tailored price signal based on the water requirements of the landscape.

Water Budget-Base Rate Structures Require a Capable Utility Billing System and Customer-Level Data

The implementation of water budget-based rate structures has been made possible by the advent of computerized utility billing systems that can incorporate specific customer-level information into a billing calculation. None of the water budget formulas or calculations studied by the research team was particularly complex or complicated. Most modern, database centered, utility billing systems can probably be adapted to incorporate water budgets without significant effort. Billing systems based on older technology may not be capable of handling water budgets. At least one utility billing software company has developed a module specifically for incorporating water budget-based billing.

The development of most water budgets examined in this study requires some customer level data such as lot size or irrigable area. Many water utilities have invested in geographical information systems that provide parcel (or customer) level data such as lot size and pervious/impervious area measurements. If these parcel level data can be linked to the customer billing system, then much of the data required to implement a landscape water budget program are already in place. Utilities that do not have a GIS may be able to use local tax assessor records to obtain lot size information and other data that could be used to develop water budgets.

Implementation Costs Vary

The costs associated with implementing a water budget-based rate structure vary tremendously depending upon the circumstances of the water agency involved. Most water budget systems have been implemented "in-house" using utility staff and limited outside expertise. Often the existing customer billing system can be modified to accommodate the new rate structure. In these situations, the implementation costs amount to the staff time required to develop the program. IRWD, San Juan Capistrano, Otay, and Centennial all implemented their water budget programs in under 12 months with most labor coming from existing staff resources.

Some water budget implementations are more expensive. The City of Boulder spent nearly three years studying and then implementing their water budget rate structure. Boulder first spent several hundred thousand dollars on a study to investigate the feasibility of water budgets. Boulder's antiquated billing system (hardware and software) was replaced as part of the water budget implementation at a cost of more than $1,000,000. It should be noted that Boulder intended to replace this system anyway and had budgeted money for the project prior to the decision to implement water budgets. Boulder also commissioned a cost of service study as part of the rate implementation which increased the overall cost of the project, but also helped to ensure its fairness to customers.

Water Budget-Based Rate Structures Spur Conservation Savings

Many of the utilities that have implemented water budget-based rate structures have experienced substantial conservation savings attributable to the rate structure and accompanying customer outreach programs. Analysis by Irvine Ranch Water District (IRWD) staff concluded that the program increased outdoor efficiency by 60 percent. In 1992 shortly after the rate structure was implemented IRWD customers irrigated approximately 3,600 acres in the service area and applied roughly 4.4 acre-feet of water per acre (52.8 inches of water applied). The typical ET requirement for turf grass in the area is 48 inches per year. In 2005 IRWD customers irrigated 11,768 acres of landscape and applied only 1.7 acre-feet of water per acre (20.4 inches). This represents a 61% reduction in the outdoor application rate since the introduction of the water budget program. An independent evaluation in 1997 documented a 37% decline in water consumption as a result of IRWD's rate structure and customer outreach (Pekelney and Chesnutt, 1997).

The same 1997 study found that two other water budget rate structures and conservation programs in California also achieved substantial conservation savings. San Juan Capistrano's water budget rate structure and customer outreach reduced consumption by 35% between the pre- and post-program periods; while the Otay Water District experienced a 20% decline in usage, after controlling for weather (Pekelney and Chesnutt, 1997).

Staff from the Centennial Water and Sanitation District in Colorado documented an average 25% reduction in demand after implementing their water budget program. Even though Centennial continues to add customers at a fairly steady rate, its demand has decreased 18 – 31% since 2003 when the water budget-based rate structure was implemented.

Most Water Budgets Are Seen As "Fair"

One of the key benefits of water budgets identified by the implementing water agencies studied for this project was their greater perceived "fairness". Most of the agency staff involved said the additional complexity of customer-specific water budgets was more than outweighed by the increased customer acceptance of the customized rate structure. Staff found that once customers understood the system they preferred to have their rates based on the characteristics of their site rather than on an arbitrary or average value. Variance programs have also been used to increase the accuracy of both indoor and outdoor water budgets since the water agency cannot know about every circumstance of each individual customer.

Landscape Penalty Rate Structures Are Less Successful

Landscape water budget systems based on a penalty rate structure, such as the one implemented by Otay in 1992 and rescinded in 2000, are more problematic than the water budget-based rate structures implemented in other cities. The Otay system relied on self reported landscape information and issued first a warning and then penalties for exceeding budgeted allotments. Although this approach resulted in 20% reduction in demand among the budgeted customers, the rate penalty system proved problematic with some customers. The system generated numerous customer complaints and the use of self-reported landscape area data allowed some customers to inflate their area estimates and defeat the intent of the system.

To be successful, water budgets must be based (to the extent possible) on measured customer level data, not self-reported information. Water budgets are also more likely to gain acceptance if they are connected with a well designed and communicated rate structure rather than a penalty structure.

Water Budgets Offer Great Potential as a Drought Response Tool

Water budgets and water budget rate structures offer water utilities powerful tools for reducing demand during drought and for monitoring customer compliance with drought restrictions. In time of shortages, water budgets allow a water provider to quickly and easily identify excess use and even penalize it if necessary. By summing all water budgets, utilities can quickly understand the amount of water likely to be required to meet customer demands in any given billing cycle. During a drought, water budgets have the potential to assist water utilities in more fairly apportioning demand reductions among customers with different needs and among different customer classes since the reference point for reductions is based on the water required by each customer in normal times.

Water budget rate structures can help with drought plan enforcement in the area of communications. The water budget rate structure, with its billing system, informs all customers on a regular basis of the required use reductions. The water bill can show each customer how much water they are allocated during the drought. This information can be developed well before the drought occurs as part of the budgeting process. This is a far more reliable and effective way to implement drought related conservation since it is pre-planned rather than improvised. The billing system is already in place and the bills can provide the public with the information needed to respond to the drought.

Another way that water budget rate structures aid with drought plan implementation is in the enforcement of mandatory demand curtailment. A simple query of the billing database can inform the utility each billing period which customers have complied with drought restrictions and remained within budget and which have not. If the higher water rates being charged are not sufficient to elicit cooperation then additional fines and penalties can be considered. This is a highly reliable system. Unlike the "water cop" approach where customers are ticketed if they happen to be observed violating the drought restrictions, a water budget drought enforcement program automatically identifies every customer who is not complying, thus enabling fair and uniform enforcement. Water enforcement patrols are costly and can only catch violators "in the act" of violating a watering restriction. A water budget, however, provides a regular and automatic check on which customers are in or out of compliance with drought response.

A Beginning

Although the water budget concept is not new, over the past five years the number of agencies implementing water budgets has grown from five to twenty-five. In the coming years it is likely that more water providers will move toward this type of conservation-oriented rate structure. This research study represents the beginning of objective analysis into the implementation and performance of water budget-based rate structures. This project was conceived, developed, and implemented in the hope of improving understanding of water budget-based rates and how they can be implemented by water utilities.

CHAPTER 1

WATER BUDGETS, RATE CONCEPTS, AND RESEARCH METHODS

INTRODUCTION

"Water budgets"—volumetric allotments of water to customers based on customer-specific characteristics and conservative resource standards—are an innovative means of improving water use efficiency. Once thought to be impractical because of technological constraints, water budgets linked with an increasing block rate structure have been implemented successfully in more than 20 utilities. As utilities develop advanced customer information systems and geographical information systems these rate structures are expected to be applied more broadly. Water budget rate structures are attractive to water agencies searching for stable revenue generation, improved customer acceptance, increased water use efficiency, augmented affordability of nondiscretionary customer consumption, and improved drought response. A growing number of utilities believe that water budget rate structures offer potential benefits to water utilities and their customers in coping with increasing water scarcity.

This research examines water budgets and their potential value to North American water utilities and the varying applications of the water budget concept that have been adapted to different conditions. The report identifies different practical approaches to water budget rate structures, the benefits and challenges of these approaches, the potential uses of water budgets during drought, and key issues involved in implementation. Numerous examples and case studies of water budget rate structure implementation are presented. Utilities interested in the concept can determine the data and technology requirements for applying water budgets and can learn from the experience of utilities that have pioneered this rate structure concept.

BACKGROUND

This research project was conceived of and developed as a way to critically examine the water budget rate structure concept. In 2002 when the project was originally proposed only five water agencies were known to be actively employing water budgets. By the end of 2006 as the project neared completion more than 25 water agencies were utilizing water budgets. This confirms the researchers' belief that water budget rate structures are an important developing trend in the water industry, well deserving of the close attention and scrutiny provided by this research effort.

The "Water Budgets and Rate Structures: Innovative Management Tools" project was funded through a tailored collaboration research grant from the American Water Works Association Research Foundation (AwwaRF); a grant from the U.S. Environmental Protection Agency, Office of Water awarded to the City of San Diego Water Department; and a consortium of water agencies and organizations including the City of Scottsdale Water Resources Department, Pinellas County Utilities, the East Bay Municipal Utility District, the California Urban Water Conservation Council, the Jordan Valley Water District, and the City of San Juan Capistrano who all provided cash and in-kind support. In-kind support was also provided by the Las Vegas Valley Water District, Otay Water District, City of Portland, City of Los Angeles Department of Water and Power, Irvine Ranch Water District, Town of Cary, City of West Jordan, and the Davis California Public Works Department.

RESEARCH METHODOLOGY

The research team used a variety of methods to obtain and analyze the information and data presented in this report. An extensive literature review was conducted using bibliographic and web-based searches. Substantial input and recommendations were also provided by the Project Advisory Committee (PAC) members. Water budget rate structure programs were identified through researcher knowledge, web searches, conversations, and interviews with water professionals. Postings on the AWWA WaterWiser web site on-line conference were also used to solicit input on agencies utilizing water budgets.

Survey instruments were developed by the research team and reviewed by the PAC. The researchers conducted numerous structured interviews with water agencies that have already implemented water budgets and water agencies interested in the concept, but not yet using water budgets. Participating water agencies provided detailed information about their water budget programs including rate schedules, memos written during the implementation process, in-house research studies, billing data analysis, customer satisfaction surveys, sample water bills, customer information pamphlets, local newspaper articles, and other pertinent information.

Additional interviews and data collection efforts were made with participating agencies that are interested in the water budget rate structure concept, but have not yet implemented the concept. This research focused on the technical ability and readiness of the agency as well as the potential barriers to implementation. Agencies also provided annual reports, consumption data, information on customer classes, and information on customer billing systems.

Case studies on implemented water budget programs were developed by the research team and reviewed by agency personnel to ensure accuracy. Case studies were chosen to be illustrative of the successes and challenges involved in implementing a water budget rate structure. The number of case studies included in this report expanded greatly as the research proceeded as it became apparent that there were substantially more water budget implementations than initially thought. Early reviewers of the report also felt that the case studies were one of the most valuable components of the project. Most case studies were provided to staff members at the water agency being studied for review. Corrections and comments received were incorporated into the final report. The researchers are extremely grateful for the time and effort spent by utility staff to ensure their stories were told accurately.

The PAC provided excellent and thoughtful input and guidance for the project throughout the entire project. Comments and recommendations from the PAC were incorporated into the final report on a regular basis as quarterly progress reports were reviewed. The authors hope this iterative process results in a final report that is as accurate, complete, and easy to read as possible.

FUNDAMENTAL CONCEPTS AND RATE MAKING

The goal of this chapter is to explain how water budgets fit into the universe of water rates and what advantages and disadvantages water budget-based rate structures might offer to water utilities. Public policy aspects of the applicability of water budget-based rates are also addressed. This chapter develops key definitions and themes that will be used throughout the report. This chapter introduces four central themes of the study:

- rate structure concepts
- water budgets
- rate structures that use water budgets
- communication about the rate structure between the utility and its customers

We first set out rate design considerations, using the guidelines suggested in the AWWA rate manuals for selecting rate structures: 1. setting rate objectives, 2. designing and evaluating rate structure alternatives, and 3. communicating with stakeholders in the ratemaking process.

RATE CONCEPTS

Rational evaluation of any rate structure—including water budget-based rate structures—should be predicated on a solid understanding of the tradition of rate making. This section provides an overview of ratemaking practices in the water industry. Some of these traditions are embedded in the AWWA rates manuals produced by the AWWA Rates and Charges subcommittee. There is also guidance and precedence available from other utility rate-setting precedents and from other international rate-setting practices.

RATE MAKING 101

Traditional ratemaking involves three discrete, logical steps (Beecher and Mann, 1991; Raftelis, 1993)[1]:

Step 1. Identify costs and the agency's revenue requirements;
Step 2. Allocate costs to types of water usage; and
Step 3. Design rates for each type of water usage to recover costs from customers.

Figure 1.1 displays these three different components of ratemaking.

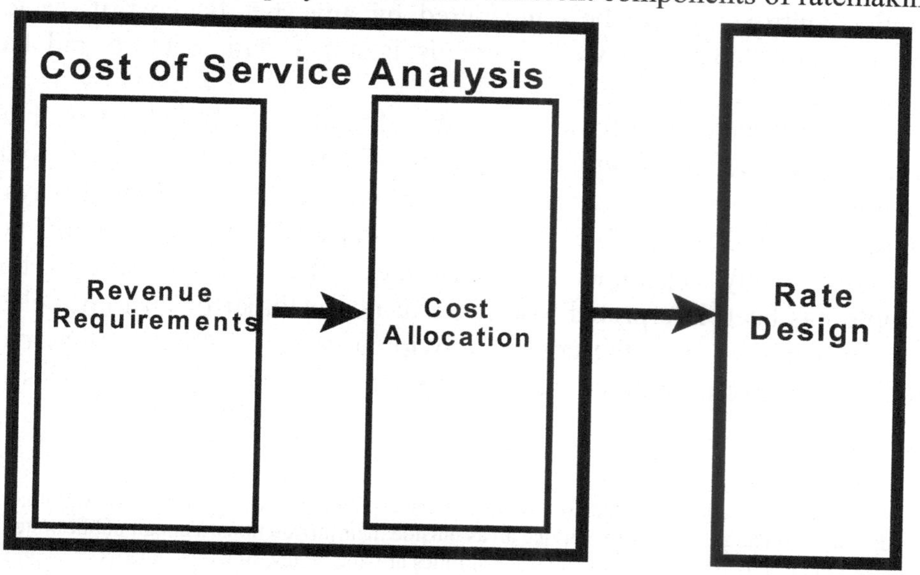

Figure 1.1 Components of ratemaking

[1] Portions of this Chapter have been adapted from the CUWCC Handbook for the Design, Evaluation, and Implementation of Conservation Rate Structures, 1996, with permission.

Revenue Requirements

To provide water service to its customers, a water agency must receive sufficient revenues to recover its costs, including operation and maintenance costs, capacity costs [represented either by depreciation allowances or by debt-related costs], customer costs, and administrative costs (American Water Works Assoc. Manual M1, 2000). The revenue adequacy objective is important because utilities that receive revenues substantially less than costs cannot, in the long run, stay in business. Adequate cost recovery is a necessary condition for maintaining a financially viable water utility.

The process of determining a utility's revenue requirement involves an examination of annual costs, including both operating expenses and capital costs. Using prudent investment standards and least cost principles, a determination is made of the revenues required to meet prudently incurred costs (Beecher and Mann, 1993). This revenue requirement represents cost commitments of the water utility—payments to bondholders, wages to employees, and other expenditures necessary to the operation of the utility. The amount of revenue that a water utility should be permitted to earn through rates is often a controversial question. Specific guidelines have been developed for the determination of revenue requirements.

Cost of Service

After revenue requirements have been established, costs are allocated among different types of water users, and then rates are designed to reflect the cost of providing water service. The concept of cost of service ratemaking evolved out of electric utilities (Breyer et. al. 2002). The idea of cost of service ratemaking can be loosely stated: rates should be designed so that users pay in water rates for the costs they impose on the utility. Though the idea may be straightforward, considerable controversy can be engendered by any specific cost-of-service analysis. The practice of accepted "cost-of-service" methods is not a static picture and has evolved with both energy and water utilities.[2]

The key legal standards that have been set are that rates should be "just and reasonable" and that rates should not be derived on an "arbitrary or capricious" basis. These Supreme Court established principles for review of rates have, in practice, been interpreted in different ways[3]. One method of establishing "just and reasonable" rates is the standard that rates should not "unduly discriminate" against any customer or customer class. In practice, this "nondiscrimination" principle has been interpreted to mean that no customer or customer class should pay significantly more (or less) than the cost of providing service to that customer or customer class.[4] To avoid undue discrimination, rate analysts strive to achieve two forms of equity:

---

[2] Though many public utilities are not subject to formal rate regulation, some aspects of public utility ratemaking may be subject to judicial review, including "outside-City" or wholesale rates for service areas outside their jurisdictions. Though public water utilities may not be subject to the same kind of rate review that private water utilities are, the review of City Councils or Boards of Directors constitute a review process that some have interpreted as "quasi-regulation".

[3] In the FPC vs. Hope case (1944) the majority opinion voiced by Justice Douglas stressed the adequacy of rates to maintain corporate credit. The dissenting opinion, voiced by Justice Jackson, stressed the need conserve the limited resource of natural gas for its most important uses—a distinctly conservation-oriented emphasis.

[4] "One of the reasons for the popularity of a cost-of-service standard of ratemaking no doubt lies in the flexibility of the standard itself….A more important disagreement is that between those who identify cost of service with some kind of average or prorated total costs, and those who identify it with differential or marginal cost." (Bonbright, 1988) p. 109.

Horizontal equity:	Users with similar costs of service face similar rates.
Vertical equity:	Users with dissimilar costs of service face dissimilar rates.

A key choice in the cost-of-service analysis is whether to distinguish costs by "class" of customer. Customer classes—homogeneous groups of customers—have been justified by similarities in service requirements and demand patterns. Both service characteristics and use patterns affect the cost of service. The implication is that customers with similar service requirements and patterns of use should be placed in the same class of service. If customer-use patterns and service requirements are similar among customers, there is little reason to have multiple rate structures; if use patterns and service characteristics vary, then the establishment of customer classifications and multiple rate structures is warranted (Costanza, 1983).

Fixed versus *Variable* Costs: Many costing methods identify costs of water service as either *fixed* or *variable* based on the characteristics of the expenditures. *Fixed* costs are expenditures that remain relatively unchanged throughout the year, irrespective of the volume of water produced. Because large up-front capital costs are required to build capacity for meeting demand, some traditional costing methods classify all system expansion costs as fixed and refer to these costs as "demand" costs. *Variable* costs, also called "commodity costs," are expenditures that vary directly with the volume of water produced or consumed; variable costs include purchased water, electricity, and chemicals. Marginal costing methods recognize that the dividing line between fixed and variable depends on the period of time used for the analysis. In the long run, fixed capital expenditures can and do change, thus becoming "variable."

Standard ratemaking methods intentionally develop pricing on the basis of "cost causation" rather than the fixed/variable cost paradigm associated with cost accounting. The intent of AWWA rate methods is to impose costs on those responsible for their incurrence rather than as a reflection of capital-intensive cost structures.

Cost of Service Analysis

Traditionally water utilities (and regulators) use cost of service studies to allocate revenue requirements according to the cost of service (including capital and operating costs) associated with different patterns of water use. A cost of service study is used to identify, for example, variations in costs caused by seasonal and daily peak demands.

Three specific activities guide the allocation of costs:

- **Cost functionalization** separates costs into functional categories, such as source development, treatment, transmission, and distribution. The calculation of costs by functional category is provided directly by the accounting system or estimated indirectly using accounting information.
- **Cost classification** assigns functional costs to service characteristics. Several approaches are possible. The base-extra-capacity method assigns functional costs to average day, maximum day, and maximum hourly usage categories. The demand-commodity method allocates functional costs to demand and commodity usage categories. Other categories are often used for customer-related costs such as billing, metering, and fire protection. An important variation classifies costs by peaking period – peak versus non-peak.
- **Cost allocation** assigns each category of costs to customer classes. For example, customer related costs are usually allocated according to the number of service connections in each class. Capacity costs are allocated differently under the base-extra-capacity and commodity-demand methods. Costs are allocated to customer classes *in proportion to* the respective demands these customers place on the utility system.

Rate Design

After revenue requirements are set and cost analyses performed, the rate analyst can begin to design the rate structure. The revenue requirement gives the design point for rates (how much revenue should be produced) and the cost analyses determine the appropriate price signal. Designing a rate structure that meets both revenue and cost allocation objectives is not, in general, enough to meet the agreed upon objectives which are typically defined early on in the ratemaking process as discussed below. There are other aims and other stakeholders in the ratemaking process. Analysts must make choices within each step. (Beecher and Mann, 1990, Raftelis, 1993).

There are several rate setting guidance documents that present a decision-theoretic approach to selecting a rate structure, including the AWWA M1 Principles of Water Rates, Fees, and Charges (2000). The AWWA M1 manual (Chapter 9) sets forth a three step selection process:

Step 1. Defining Goals and Objectives of the Rate Structure
Step 2. Evaluating Available Alternatives in Meeting Objectives
Step 3. Understanding and Communicating Potential Effects on Customers

Step 1. Defining Goals and Objectives. The first step is defining the objectives that a rate structure is supposed to accomplish. These objectives both guide the design of a rate structure and form the standard against which alternative rates can be evaluated. The M1 manual lists potential rate objectives but provides limited discussion. Additional review of rate objectives can be found in Bonbright (1988), Chesnutt et al. (1996), or Raftelis (1993). In a public involvement process, these objectives are often elicited. Typical objectives might include:

- Revenue Effects—Revenue sufficiency, net revenue stability.
- Consumption Efficiency— Water rates send price signals to customers about the worth of reducing or augmenting water consumption. This includes load management (efficient utilization of existing facilities).
- Resource Conservation— From the consumptive decisions of customers, agencies are sent signals about the worth of additional or improved water service. Rates should also balance costs that do not directly appear on utility financial books (externalities.) For example, water supply development costs likely understate the true economic costs of these supplies in that development costs do not reflect environmental and social costs incurred outside a market context.
- Consistency with cost-of-service principles—Users whose service requirements impose costs on the utility should pay for those costs.
- Administrative cost coverage—Utilities incur direct costs in administering rate structures.
- Institutional support—Rates often must be approved by a city council, board, or public utility commission
- Affordability—Low income customers may have difficulty paying their water bills. Since some amount of water is a basic human need, many water utilities attempt to address the broader social goal of providing affordable water service.
- Public acceptance—Water rate structures must be accepted by customers.
- Equity—Perceived "fairness" of the rate structure is a key determinant of public acceptance. There is more than one valid definition of "fairness".

Step 2. Evaluation of Rate Alternatives. The second step is rate evaluation where alternative rate structures are evaluated to determine how well they stack up against the objectives. What are the available rate structure alternatives and how should they be evaluated? One way to divide the universe of rate structures is to distinguish fixed charges (fees, taxes, or some constant dollar amount) from volumetric rates (price per unit volume). A *rate structure* can be composed of both fixed charges and volumetric rates.

When designing rates, utilities must determine whether to recover costs through *rates*, which vary with usage, or *fixed charges*, which do not vary with usage. The revenue collected through commodity charges varies with the amount of water used by customers; revenues from fixed charges are not sensitive to use. Examples of fixed charges include service charges and "readiness-to-serve" charges. Rates can send a message that consumers should conserve water; fixed charges provide no incentive to reduce water use. A common practice in rate design is to allocate costs that are sensitive to usage (such as pumping costs) to a volumetric rate, while allocating costs that are not based on usage (such as connection costs or customer costs) to the fixed charges. This research does not focus on the determination of fixed charges or on the allocation of costs between fixed charges and variable rates. Instead, this report will assume that a determination of the revenue required from volumetric rates has been made for the customer class in question. That is, we will assume that the revenue requirements from sales of water are known.

To understand water budget-based rates, two types of volumetric rates will be distinguished. Figure 1.2 depicts the simplest type of volumetric rate--a uniform rate where the same unit price (dollars per gallon or cubic foot) applies to all water use. A variation to a single uniform rate for all customers is uniform rates by customer class, where the rate varies by customer class but not within each class. Uniform rates are relatively simple, easy for customers to understand, and typically involve low administrative costs. Uniform rates are generally compatible with prevailing notions of equity and fairness. Uniform rates also recognize that certain unit costs of water provision (for example, treatment) remain relatively constant with increasing water use.

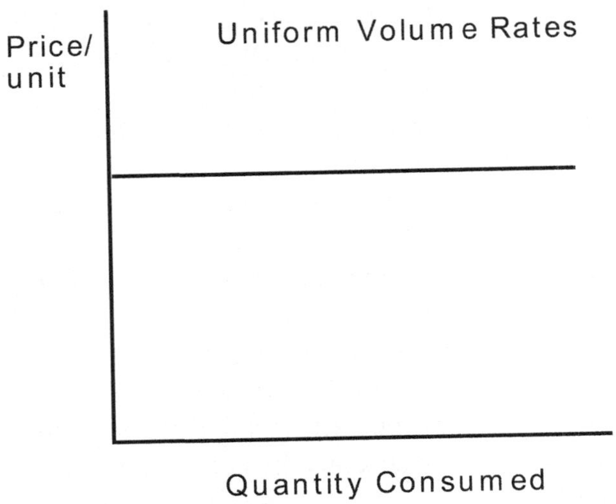

Figure 1.2 Uniform rate structure

Figure 1.3 depicts an increasing-block rate structure, the applicable unit price increases with the higher use blocks or tiers. The conceptual basis for the increasing-block rate as a conservation rate form is the belief that unit capacity costs associated with system expansion will increase with more costly supply sources. The argument in favor of increasing block rates lies not in some intrinsic merit to charging different prices for different units of water consumed at the same point in time. Rather, the strongest argument for increasing block rates is practical: Under a revenue constraint, increasing block rates allow agencies to charge efficient prices (equal or close to the long run marginal cost) to more customers than would otherwise be possible.[5] Price increases associated with inclining blocks may also be used to reflect the costs of peak period use as much as the increasing costs of supply.

[5] If the cost structure of the water industry is characterized by increasing costs, an efficient price would reflect the higher (forward-looking) costs. Reflecting this higher cost in a price on all units of consumption would result in excess revenues—revenues collected on all units of consumption (based on a higher forward looking price) would exceed the revenue required for current costs (incurred from a mix of older costs that are lower and newer costs that are higher.) Increasing block rates collect less revenue from the first units of consumption while charging a higher and more efficient price for later units of consumption. In this way, they can send a more efficient price signal to more customers without over-earning.

In actual practice, increasing block rate designs often have precious little to do with marginal cost pricing. Many utilities simply do not have a good idea of what marginal costs they face, and even fewer have consciously attempted to tie their pricing to these costs. In spite of the theoretical underpinnings, the reality is that most inclining block rates attempt to provide a politically acceptable conservation price signal.

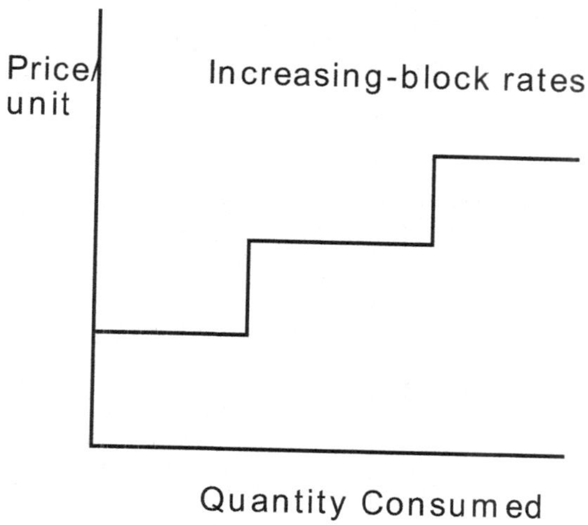

Figure 1.3 Increasing block rate structure

Other variations of volumetric rates include seasonal rates (where higher prices are charged in peak seasons when costs are higher and/or supply reliability is lower) and drought pricing (where higher prices are charged during periods of scarcity to more efficiently allocate water in times of shortage.)

Once a range of alternative rate structures has been defined, options can be evaluated against the selected criteria. Figure 1.4 displays one depiction for how rate structures can be evaluated. Water utilities entertaining water budget-based rate structures may need to allocate more time and resources to rate evaluation to obtain a better idea of their relative costs and benefits.

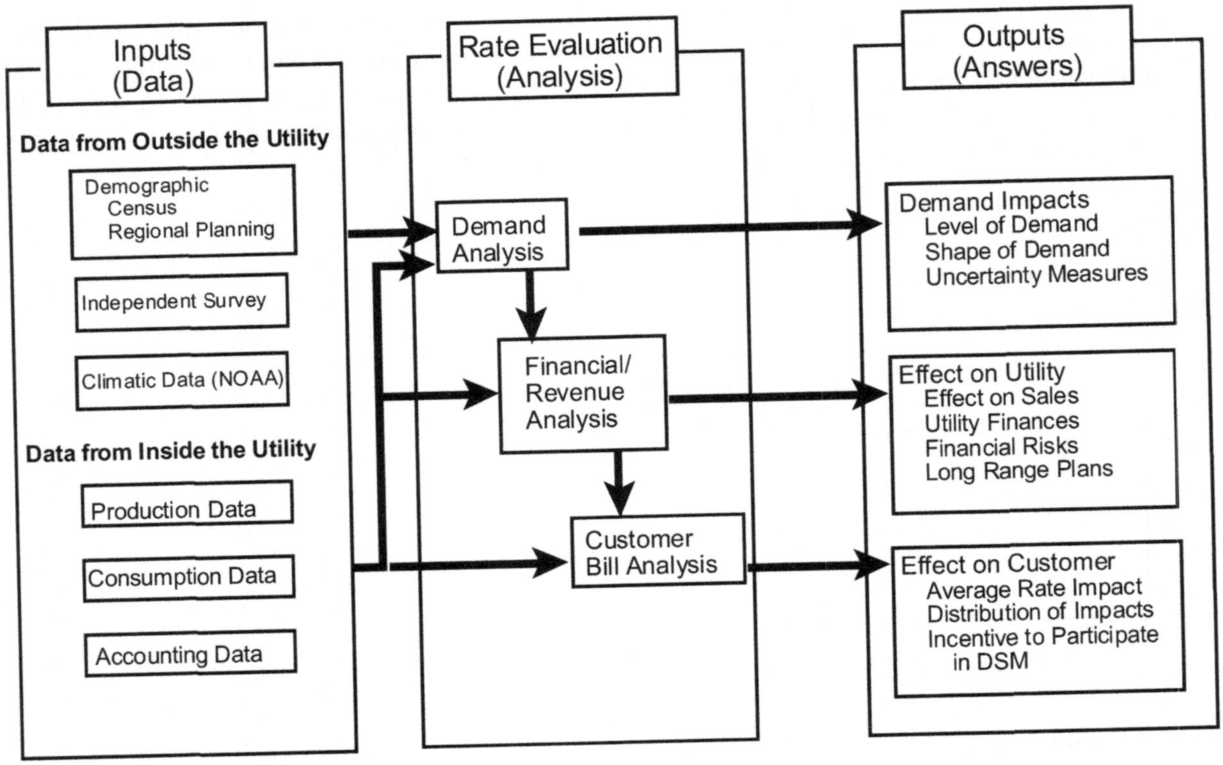

Figure 1.4 Rate evaluation

Step 3. Communication. The third step is understanding and communicating the potential effects of a rate structure on customers. Increased attention has been placed within the water industry to customer communication imperatives—improving customer understanding of both the rate structure that generates a customer bill and the service paid for by their water bills. Recent examples of this concern include *Avoiding Rate Shock (2005)* and *Thinking Outside the Bill (2005*). Water budget-based rate structures require additional effort to communicate the definition of efficient water use defined by the water budget.

WHAT IS A WATER BUDGET?

Water budgets (also known as water allocations) are a familiar water management tool to many agricultural water customers, horticulturalists, and other green industry (landscape industry) professionals. A water budget is a formal definition of the quantity of water that would be required by an efficient level of water use. Exactly what constitutes an "efficient level" of use (particularly as it relates to outdoor irrigation) is often a community decision and must be based upon public expectations as well as water supply realities. Water providers tend to define water budgets in more specific terms so that their customers can easily understand how the budget (or allocation as they are often called) applies. Table 1.1 provides a sampling of different definitions or explanations of water budgets/allocations.

Table 1.1 Utility definitions of water budget/allocations

Location	Water Budget/Allocation Definition
City of Boulder, CO	A water budget is the amount of water you are expected to need for a specific month. Each customer's water budget will be different based on their water needs. Water budgets may vary monthly based on seasonal outdoor watering needs.
Irvine Ranch Water District, CA	IRWD uses a five-tiered rate structure which rewards conservation. Water allocations are based on number of residents, landscape square footage and actual daily weather and evapotranspiration (ET) data for your area. Evapotranspiration is the amount of water that travels through turf grass, and therefore the amount of water required for turf grass to be healthy and attractive.
San Juan Capistrano, CA	Allocations are made up of a combination of an Indoor Allocation and an Outdoor Allocation. The sum of the Indoor and Outdoor Allocation serves as the Tier I Allocation. Use above this level may be penalized by a higher commodity rate. The indoor allocation is applied equally to each month. The Outdoor Allocation is calculated each billing period (monthly) based upon local weather data for the actual days in the billing period. The Outdoor Allocation is calculated taking into account the Evapotranspiration Rate (ETo), Crop Coefficient for turf grass, Management Coefficient and Effective Rainfall.
Centennial Water and Sanitation District, CO	The water budget approach to rate structuring provides an amount of water tailored to individual customers based on actual needs. Each customer's water budget is based on estimates of indoor and outdoor needs. The budget system provides water quantities appropriate to each user before a higher rate is applied. It is also a means of encouraging water conservation by rewarding customers who stay within the budget amounts.
Santa Rosa, CA	The Billing Period Water Budget is the amount of water, in gallons, needed to irrigate the specific area of landscape served by the Dedicated Irrigation Account for the specific days covered by the billing period.

Landscape water budgets are the most common form of a water budget. A landscape water budget is typically a volume of water that is calculated from two fundamental parameters: the landscape size (usually in square feet) and the water requirement of plants in that area which is typically represented by the evapotranspiration rate (ET or ETo).[6] Additional factors such as irrigation efficiency and landscape material (i.e. plants with differing water requirements) may also factor in. Detailed formulations of different landscape water budget calculations are presented later in this report.

Indoor water budgets for residential customers generally fall into two categories: (1) fixed budget; or (2) per capita budget.

[6] There are a number of methods for calculating ET. A key difference is that many methods do not include precipitation in the calculation. ETo is typically a measure of ET that does not include precipitation.

A fixed budget is a set volume of water, typically 5 – 7 kgal/month[7] for a single-family customer and 3 – 6 kgal/month for a multi-family customer, determined by the water provider. A per capita budget is tailored for each customer based on the number of people living in the residence. For example a family of 4 with a budget of 70 gallons per capita per day (gcd) would receive 8.5 kgal per month for their indoor budget.

Indoor water budgets for non-residential customers are typically determined using two different methods: (1) average consumption; and (2) usage in customer class. The average consumption method sets a water budget based on historic use at the site. This method customizes the budget for each customer, but also potentially grandfathers in inefficient water use by setting higher budgets for customers who have historically wasted water and lower budgets for customers that have been efficient historically. The usage in customer class method requires that all non-residential customers be identified by classification system where office buildings, restaurants, hotels, schools, churches, day care facilities, etc. are explicitly identified. The water budget for each customer is then based on the customer class, a benchmark usage value for that class, and some measurement of size (e.g. number of: workers, meals served, hotel rooms, students, parishioners, etc.). This method is more complex and data intensive, but has potential for sending an effective price signal to customers who are inefficient within their customer class. Systems that have not collected data on land uses and customer classification may find this particular methodology cost prohibitive to implement.

Further explanations and examples of the methods for developing indoor and outdoor water budgets are presented later in this report.

WHAT ARE WATER BUDGET-BASED RATES?

Water budget-based water rates—also known as individualized, goal-based, and customer specific rates—are block rates where the block is defined by using one or more customer characteristics. For example, these characteristics could include the number of persons per household, lot size, or the evapotranspiration requirements of a customer's landscape. Typically, a lower rate is set for levels of water use that are less than the block definition and a higher rate is set for levels of water use that are greater than the block definition.

Water budget-based rate structures can be thought of as an increasing block rate structure where the block definition derives from a water budget that defines an efficient level of water use for each customer. A diagram of a simple water budget rate structure is shown in Figure 1.5.

[7] kgal = kilo-gallons = 1,000 gallons = 134 cubic feet

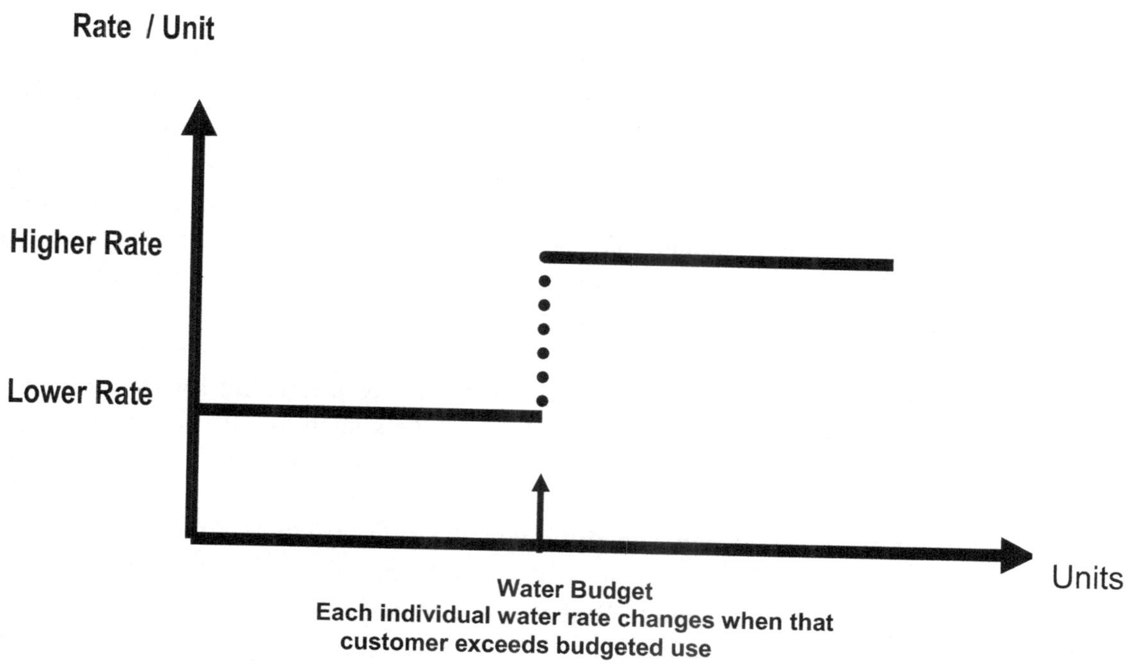

Figure 1.5 Water budget rate structure

WHY USE WATER BUDGET-BASED RATES?

This report seeks to outline the potential costs and benefits of water budget-based rate structures. Using water budget-based rates requires customer-specific data in order to define the rate structure, additional staff time to design and validate the rate structure, and an additional burden to communicate a more complicated rate structure. Such an undertaking necessarily involves potentially significant additional effort; why would anyone want to undergo such an exercise? Water budget-based rates may not be suited for every situation. They have, however, been successfully implemented in several areas and been widely viewed as successful. It is important to understand the conditions that led to the successful adoption of these rates in order to understand their potential benefits and risks.

Water budget-based water rates have been implemented in communities like IRWD, East Bay Municipal Utility District, Santa Barbara, and Capistrano Valley, California facing limited water supplies or shortages. Field evidence from their implementation sheds some insight on the advantages and disadvantages of water budget-based rates. On the plus side, these rates have been considered an equitable way to share limited water supplies while preserving some amount of customer choice. Indeed, the largest strength identified with water budget-based rates—by both customers and water utility staff—is their perceived fairness (Pekelney and Chesnutt, 1997). If the basis of the water budget-based rate and the supply constraint situation are communicated well, then the rate is seen as an intrinsically more equitable way of charging water rates than rate structures that do not include individual customer characteristics. Water utilities that have implemented water budget-based rate structures have developed closer working relationships with their customers while coping with water supply constraints.

The implementation of new information technologies (especially AMR and geospatial parcel measurement applications) hold the prospect of broader, less costly, application of water budget based rates. Many water agencies have already developed information systems that contain sufficient information and data for the development of water budgets. Modern utility billing systems that are database-driven are often more capable and easily adapted than older billing technologies. Typical water budget rate calculations are not particularly complex and often only require the inclusion of a few additional fields and a calculation or two in a utility billing database. The water budget toolbox section of this report focuses on information that can help agencies realize opportunities to utilizing water budgets. The implementation of new technology such as GIS and integrated agency databases can make it feasible for utilities to use water budgets. The subject of utilizing information technologies such as GIS to support water rates and other related projects is ripe for future AwwaRF research projects.

The disadvantages of water budget-based rates include potentially higher implementation costs including a billing system that can accommodate the rate structure and customer level information that may not be readily available. More upfront effort to implement the rate structure may be required of utility personnel. Water budget-based rates are a departure from the billing methods most customers are familiar with and customer service representatives must be prepared to handle questions and problems that may arise. Successful water budget implementation requires a level of commitment from decision makers, staff, and citizens. Many agencies have opted to implement water budgets for irrigation only accounts to gain experience with the concept before tackling a utility-wide program.

CHAPTER 2

LITERATURE REVIEW – RATES AND DEMAND MANAGEMENT

Over the past century, lush landscapes, handsome public water features, and a seemingly limitless supply of water have been used effectively to attract new businesses and residents to an area. In many areas, utility customers tend to view water as an abundant and inexpensive commodity, and providers encourage this perception by offering large volumes of water at inexpensive rates. Yet many of the fastest growing areas in the United States are located in the South (Florida, Georgia, the Carolinas) and West (Arizona, California, Colorado) where available water supplies are limited. Explosive growth, inefficient use of water, increased regulatory constraints, and droughts have now contributed to the need for these communities to consider ways to increase the efficiency of water use. Education, efficient technology, and pricing mechanisms are the most commonly used tools adopted by utilities in their efforts to reduce water demands. In many areas of North America, controversy surrounding the need to protect supply sources, optimize conjunctive uses, and promote more efficient use of water has evaporated in recent years. Agreement about the best way to promote water efficiency has been harder to achieve.

Utilities often grapple with the need to change or adapt their rate structures to ensure an adequate supply of water, renew and replace aging infrastructure, ensure regulatory compliance, and meet the needs of their customers in a fair and equitable manner. Because water is fundamental for human life, utilities try to ensure affordable water rates over a long period of time. The high cost of both new water supply projects and infrastructure renewal has awakened many utilities to the realization that stretching existing supplies and treatment capacity through conservation is a viable low-cost option compared with raising revenues through increased water sales and system expansion.

WATER BUDGETS

Because rates are known to be an effective tool for demand side management, utilities have continually sought to implement new rate structures that provide an effective conservation signal (Vickers, 2001). Since the early 1990s, water budgets – volumetric allotments of water to customers based on conservative resource standards – coupled with an increasing block rate structure have been considered among the most effective rates for conveying efficiency via meaningful price signals (Chesnutt and Mayer, 2004). Yet because of the data requirements and technical complications, these rate structures have only recently begun to be adopted on a more widespread basis.

Water budgets are perhaps the most important and innovative developments in water billing in a generation. Yet research on these policies remains scattered and anecdotal and few studies have documented the success of water budget programs (Pekelney and Chesnutt, 1997; Bamzai, Silva, and Ash, 1997; Chesnutt and Beecher, 1998). Others have explored financial and economic aspects of individualized rate structures at a theoretical level (Collinge, 1996; Teodoro, 2002). However, no study has explained the lessons learned about water budgets from a rigorous, comprehensive scientific approach in practical terms understood by water utility financial managers. No recent study has critically examined the benefits and costs associated with water budgets, nor the potential for this structure to be used to help manage demand and drought-induced supply shortages.

FUNDAMENTALS OF RATE MAKING

M1 Manual

The American Water Works Association Manual M1, Principles of Water Rates, Fees, and Charges (5th ed.), (AWWA 2000) is an effort by the AWWA to consolidate the information made available in several of its previously published manuals (AWWA M26, M34, M35). The goal of AWWA Manual M1 is to provide information and guidelines to both governmental and investor-owned utilities on a diverse group of issues ranging from the determination of revenue requirements, selection of rate structures and pricing alternatives, identification of customer groups, to consideration of public policy and legal issues. The numerous examples used throughout the manual are not intended to be an endorsement of any particular methodology. Rather, they are a demonstration of ways in which utilities can continue to serve the needs of their customers and have adequate income when faced with growing demand on supplies, asset renewal, equipment replacement, etc.

Water budgets are conspicuously absent from the current M1 Manual. The authors address many of the related issues and concepts including demand side management, increasing block rates, weather factors, and seasonal rates. Individualized rate surcharges are briefly discussed in the chapter on "Drought Pricing". The authors also address some inequities inherent in an increasing block rate structure that are sought to be corrected with water budgets.

> *"Increasing block rates are not a one-size-fits all solution. System-wide application of a single increasing block rate structure is likely to result in cost-of-service inequities, especially in commercial and industrial customers with relatively constant consumption patterns (low peak demands but high total usage). These customers may not impose costs on a water system proportional to the cost implied by increasing block rates. Additionally, assigning large price increases on these customer classes, known to have the most price-elastic demand, can make it difficult to predict decreases in consumption. A single system-wide increasing block rate design applied to a customer base with diverse consumption patterns is more difficult to justify on a cost-of-service basis than increasing block rates targeted to specific customer classes with relatively homogenous consumption patterns." (AWWA 2000, pp. 99-100)*

The final sentence in particular relates well to the concept of water budgets. The M1 authors imply that block rates applied on a customer class by class basis (i.e. different block rates are applied to residential and CII customers) are easier to justify on a cost-of-service basis. This is a first step in the customization of water rates to the needs of each customer class. Water budgets take this idea of customization several steps farther with the application of a block rate structure tailored by an individualized budget to the utility defined water requirements of each customer. The quote above suggests that as increasing block rates become more targeted they are also more easily justified on a "cost-of-service" basis than a single system-wide increasing block rate. Although the M1 Manual doesn't mention water budgets, the implication is that as rates become more customized and tailored to individual customer they are *easier* to justify. Since water budget rate structures are often further tailored to the customer level, they should also be justifiable on a cost-of-service basis.

Cost of Service

According to the M1 Manual, "the basic premise in establishing adequate rate schedules that are equitable to different customers is that rates should reflect the cost of providing water service" (AWWA 2000). There are two fundamental methods for calculating cost of service according to M1, the Base-Extra Capacity Method and the Commodity-Demand Method. It is unnecessary and beyond the scope of this project to present a full explanation of the intricacies of cost of service calculations. However, it is within the scope of this project to examine and explain how water budget rate structures can be implemented to satisfy cost of service requirements. The concept of cost of service can be drawn with a broad brush and a wide variety of applications are evident in water rate structures across North America. The research team could find no report or publication that specifically identified cases where water budget rate structures failed to meet cost of service criteria. Given that more than 20 utilities are actively implementing some form of these rate structures, it appears that they can be adapted to meet these important criteria. However, the ultimate proof of a water budget rate structure's consistency with cost-of-service is not in the absence of studies suggesting the contrary but rather in the very nature of the rate design.

Price Elasticity

The relationship between the retail price of water and consumption has been thoroughly explored because of its implications for water planners and providers, and its importance in determining rate structures and levels. A basic economic assumption is that water price and consumption are inversely related and a convenient way to quantify this relationship is through the concept of price elasticity. Elasticity is defined as the percentage change in quantity consumed for each percent change in price. An elasticity of negative one implies that a one percent change in price results in a one percent decline in quantity consumed. Price elasticity measures the sensitivity of water use relative to changes in the price of water, after controlling for various influential factors such as weather and income (AWWA 2000). In general, the price elasticity of water is negative because theoretically, as the price of water increases, water use decreases. A price elasticity with an absolute value between zero and one is considered inelastic, or relatively unresponsive to rate change. An absolute value greater than one is considered elastic. A product's elasticity depends on how many uses it has, the quantity and availability of its substitutes, and its relative importance within the consumer's overall budget (Schlette and Kemp 1991).

Numerous studies have been conducted on this topic and a summary of some of the more relevant studies is provided in Table 2.1. A review of more than 50 water demand studies concluded that the typical range for price elasticity is -0.2 to -0.4 for total residential demand, between –0.7 and –1.6 for residential outdoor demand, and between –0.06 and –0.8 for residential indoor demand (Planning and Management Consultants 1984). Overall, water demand over the past 50 years revealed price elasticities that ranged from –0.01 to –1.38 (Hanemann 1998). The variability in elasticity estimates for residential demand was investigated by Espey, Espey, and Shaw (1997). A meta-analysis was performed on 124 elasticity estimates from 24 journal articles from 1967 to 1993. The rate structure, evapotranspiration rates, rainfall, and season were all found to significantly affect price elasticity. Income was also found to be an important explanatory variable, although it was not found to significantly affect price elasticity in their study. Residential versus commercial demand, as well as long-run versus short-run price responsiveness, were also deemed influential. On the other hand, population density, household size, and temperature were not found to significantly affect price elasticity.

Table 2.1 Chronological summary of selected price and elasticity studies

Researcher	Price Elasticity	Notes
Howe and Linaweaver (1967)	-0.231	*21 areas in US:* Residential indoor use.
	-0.703	*10 areas in Western US:* Residential outdoor use.
	-1.57	*11 areas in Eastern US:* Residential outdoor use.
Gibbs (1978)	-0.51	*Miami, FL:* Using marginal price.
	-0.62	*Miami, FL:* Using average price.
Camp (1978)	-0.03 to -0.29	*10 Northern MS cities:* Linear equation.
	-0.35 to -0.40	*10 Northern MS cities:* Logarithmic equation.
Danielson (1979)	-0.27	*Raleigh, NC:* Using disaggregated data for total residential demand.
	-0.305	*Raleigh, NC:* Using disaggregated data for winter demand.
	-1.38	*Raleigh, NC:* Using disaggregated data for summer demand.
Billings and Agthe (1980)	-0.27 to -0.61	*Tucson, AZ:* Using two price variables and increasing block rates.
Carver and Boland (1980)	-0.1	*Washington, DC:* Short-term residential demand.
Howe (1982)	-0.06	*21 areas in US:* Residential indoor use.
	-0.568	*10 areas in Western US:* Total summer demand.
	-0.427	*11 areas in Eastern US:* Total summer demand.
Planning and Management Consultants (1984)	-0.2 to -0.4	*National:* Total residential water use, reviewed over 27 water demand studies.
	-0.7 to -1.6	*National:* Outdoor residential water use, reviewed over 5 water demand studies.
	-0.06 to -0.8	*National:* Indoor residential water use, reviewed over 5 water demand studies.
Billings and Day (1989)	-0.72	*3 utilities around Tucson, AZ:* Increasing block rate and service charges.
Schneider and Whitlatch (1991)	-0.262	*Columbus, OH:* Long-term residential.
	-0.119	*Columbus, OH:* Short-term residential.
Hewitt and Hanemann (1995)	-1.57 to -1.63	*Denton, TX:* Discrete/continuous choice model, block rate pricing of residential demand, may have been influenced by summer irrigation.
Michelsen, et. al. (1998)	+.02 to -0.23	*National:* Seasonal analysis. (+.02 in winter, -0.18 in fall, -0.13 in spring, -0.23 in summer, -0.15 annually)
Hanemann (1998)	-0.01 to -1.38	*National:* Reviewed municipal and industrial water demand studies from 1951 to 1991.
Goodman (1999)	-0.7	*57 US cities:* Extrapolates from single-family to multi-family sector.
Pint (1999)	-0.04 to -0.47	*Alameda County, CA:* Total summer demand, steeply increasing block rates during drought.
	-0.07 to -1.24	*Alameda County, CA:* Total winter demand, steeply increasing block rates during drought.
Cavanagh, Hanemann, and Stavins (2001)	-1.00*	*11 cities in US and Canada:* Using block price structure.
	-0.19*	*11 cities in US and Canada:* Using uniform marginal price structure.
Mayer, et. al. (2004)	-0.27	*12 US cities:* Multi-family indoor use
	-0.07 to -0.16	*Multi family:* Submetered indoor use. 26 properties in 3 different geographic regions.
Whitcomb (2005)	-0.39 to -0.84	*16 Florida cities:* Single family long-run elasticity ranges from four wealth profiles.

* *Price coefficients (measure household sensitivity to price), not elasticities.*

Most studies have found that indoor demand is less elastic than outdoor demand. Howe and Linaweaver's study in 1967 was the first study to separate residential demand into two components of indoor and outdoor use. This landmark study strongly indicated that price has a significant impact on water consumption, a relationship that had not been found in other studies that did not separate indoor and outdoor consumption. In this study, thirty-nine areas in the US were examined. It was determined that indoor use, estimated from winter consumption, was relatively inelastic (-0.23). Outdoor demand was more elastic, especially in the humid eastern climates compared to the dry West. Howe refined the study in later years, and found an indoor elasticity of –0.06 (1982). Danielson (1979) found a higher indoor elasticity of –0.30. Both studies confirmed that indoor use was less elastic than outdoor use.

Indoor use is less elastic than outdoor use because indoor uses tend to be fundamental to human survival and lifestyle. There are few substitutes for indoor water, beyond bottled water for drinking, and the many essential hygienic uses contribute to its inelasticity. Outdoor water use decisions may be based on consumer choice not need. (i.e. it is less imperative that a car be washed or a lawn watered). It is this discretionary water use that makes it more elastic. In addition, water bills usually represent a small portion of an overall residential budget. The average US household served by a public water and wastewater system paid approximately $39 per month in 1999 for the water portion of the service, which represented 1.6% of the household's income (Rubin, 2005). The actual commodity charges for water (excluding fixed charges, waste water, etc.) were found to be less than $16 per month. Furthermore, the typical resident is unaware of how much water they are using for any given purpose, thus contributing to the price inelasticity (Cavanagh, Hanemann, and Stavins, 2001).

One of the most contentious subjects in estimating price elasticities revolves around the price variable. The price variable is a complex subject because often, monthly water bills are not straightforward calculations. Most studies have used either a marginal or average price. Marginal price is based on the cost of providing the next unit of water, whereas average price simply divides the total cost by the total water use. Gibbs (1978) was the first to promote the use of marginal rather than average price in determining elasticity measures. Billings and Agthe (1980) employ marginal price and a difference price variable. However, price perception tests found that customers respond to average price rather than marginal price in electricity bills and water bills (Nieswiadomy 1992).

Since most water budget rate structures have been designed around landscape water budgets, the concept of price elasticity becomes highly relevant. The decision on where to set block prices can significantly influence demand. The goal of many landscape water budget programs has been to eliminate outdoor water waste, but not necessarily to reduce what is considered legitimate irrigation of turf and plants. The construction of the water budget, design of the rate structure, and the setting of the various block rates entwine and each becomes a significant factor in shaping customer response to the rate program.

OTHER PRICING AND RATE STRUCTURE ISSUES.

True Cost Pricing

It has been suggested that the practice of billing customers simply for recovering the cost of providing and distributing water does not reflect the true price (or value) of water and consequently inefficient and wasteful practices have been encouraged. This "pure" economics argument also contends that the use of complex rate structures have little or no impact on water consumption. If water customers are considered pure "economic beings" who respond only to price, then any type of rate structure should be both simple and a true reflection of the cost of providing that water (Jordan and Albani 1998). Others suggest that "water overuse is an artifact of water under-pricing, and once prices embed the appropriate social values, the need to penalize large water users evaporates" (Griffin 2001).

As water conservation becomes a priority in many communities, utilities are reluctant to rely solely on price in their efforts to curtail water use. There are economic and social implications that must be considered when price alone is used to affect water consumption. There is concern amongst utilities that "a reduction in water use that is proportionally greater than the rise in price will result in reduced revenue to the utility, and in some systems may create budgetary problems" (Cummings, et. al. 2004). There is concern also that when rates are higher for all uses there is the potential for creating a financial hardship for low and fixed-income users (AWWA 2005).

Conservation and Rate Structure Forms

Considerable attention has been given to evaluating various rate structures for their ability to provide adequate utility revenues while maintaining reasonable rates and promoting water conservation (Western Resource Advocates 2005), (Whitcomb 2005). The four most common rate structures are: uniform, decreasing block rate, seasonal, and increasing block rate (NC League of Municipalities 2005), (Western Resource Advocates 2005). Using a 2002 survey of more than 150 cities the breakdown of rate structures shown in Table 2.2 was calculated (Raftelis 2002):

Table 2.2 Prevalence of rate structures in United States

Rate Structure	Residential	Non-Residential
Uniform	37.2%	45.9%
Increasing block	29.1%	17.6%
Decreasing block	30.4%	33.1%
Other	3.4%	3.4%

Adapted from Raftelis 2002.

Some form of block rate structure (increasing or decreasing) is the most commonly used, but the uniform rate structure remains the single most popular approach. In general utilities appear more willing to apply a block rate structure to residential customers, preferring a uniform rate structure for non-residential customers. In practice, the use of uniform rate structures for non-residential classes may be less a matter of utility preference but rather recognition that the diversity of land uses within traditional customer classes may frustrate the efficacy and equity of block designs.

There is general agreement that a decreasing block rate structure encourages more use while uniform rates, seasonal rates, and increasing block rates offer potential for promoting water use efficiency. The efficacy of seasonal rates in promoting water conservation depends in large part on seasonal variations in water use and/or population fluctuations. Seasonal uses, such as irrigation, are typically discretionary.[8]

Increasing block rates have considerable potential for encouraging conservation; however the extent to which this rate structure has been adopted may be a reflection of the degree to which municipalities value conservation and the extent to which they have experienced drought or other stresses on the available water supply. In the North Carolina Water and Sewer Rates and Rate Structures report (2005) of 276 utilities, only 49, or 18% of the utilities, used an increasing rate structure and more than half of those were utilities with populations of less than 2,500 residents. A similar study by Western Resource Advocates (2004) found that of the 14 utilities studied on Colorado's Western Slope, six utilities (43%) used an increasing rate structure. Among the 12 utilities studied along the drier, more populous Front Range 67% used an increasing block rate structure.

A study in Florida by Whitcomb (2005) found that 10 out of 16 utilities used an increasing block rate structure. This study concluded that pricing can be an effective tool for managing water resources even though customers are often ignorant of both the prices paid for water and the amount of water consumed for different end uses. The author concluded that in spite of this contradiction, customers are aware and concerned about the total dollar amount of the bill and adapt their water use behavior accordingly.

IMPLEMENTING A CONSERVATION ORIENTED RATE STRUCTURE

Moving to a rate structure that promotes efficient use requires a thorough analysis of various conservation rate structures and benefits they can provide to the utility and its customers. Many social and economic issues must be addressed when changing from a rate structure that encourages water use to one that promotes conservation (Wang, et. al. 2005).

As efficient use of water becomes the long-term goal of many utilities, planning becomes an essential part of reaching that goal. Questions of how much water must be saved, how to achieve those savings, how to measure the success of conservation programs, and how to maintain a healthy revenue base may require strategies significantly different from more traditional customer service programs. Water utilities considering a water conservation program have a variety of tools that will help to ensure their water conservation program is a success and revenue impacts are mitigated. Water budget rate structures are listed as one of these tools (AWWA 2006).

A 1995 Washington State Department of Health report explored the potential of conservation oriented rate structures in promoting water efficiency within the state (Washington State DOH 1995). The authors examined a variety of rate structures from the more typical uniform, declining and inverted rates to those that are less common such as time of use, excess use, and individualized goal billing rates which are often akin to water budget rate structures. The authors concluded that conservation rate structures are an effective way to promote water efficiency and can also provide stable utility revenue generation, even if demand is reduced. In the Washington study, goal billing rates had been implemented in 25 percent of water agencies surveyed. While goal billing differs from the water budget rate structures examined in this study in that the goals are seldom customized to the individual customer level, the similarities are

[8] Some researchers dispute the conservation impact of a uniform seasonal rate structure, arguing that there is an insufficient price signal for promoting conservation (Western Resource Advocates 2005).

notable and this research suggests that a wide number of agencies in the Washington area have actively pursued this approach.

The interplay of aging infrastructure, population growth, and drought are also factors utilities must consider when moving to a conservation oriented rate structure (Barta 2004). Examining the experiences of water providers and the green industry (i.e. landscape workers and professionals) during and after the Colorado drought of 2002-03, Barta explored the impact of outreach and education, policy and regulation, and conservation-oriented pricing structures for promoting water conservation. Barta concluded that there is a lack of peer reviewed studies on non-price conservation programs and that the failure to adequately account for the impacts of non-price programs could result in over-estimates of savings from price programs.

Water Budget Rate Structures

Water budget rate structures have been successfully implemented by mid-sized utilities since the early 1990s. Until 2000, only a handful of agencies had adopted the water budget concept, but over the past five years a growing number has decided that water budgets (particularly for dedicated landscape customers) are an effective way to equitably promote water efficiency (Chesnutt and Mayer, 2004). Rather than lump customers together in large classes such as residential or commercial, water budgets - also known as individualized rates, are based on individual characteristics of customers and customer classes. Those characteristics may include the number of household members, irrigated area, type of landscape, regional evapotranspiration rate, or some combination thereof. Many have recognized the burden that developing and implementing water budgets may create for utilities, but studies have shown that "if the basis of the individualized rate is communicated well, then the rate is seen as an intrinsically more equitable way of charging water rates than rate structures that do not include individual customer characteristics" (Chesnutt and Pekelney 2002).

A 1997 evaluation of landscape water budget programs in three California water agencies found a "great…satisfaction by…water agency staff and board members with…water-budget-based rate structures as an effective and fair water conservation tool" even though many staff members began as skeptics (Pekelney and Chesnutt 1997). Utility customers surveyed as part of the same study reported favorably about conservation program performance reporting it as, "clear and understandable, responsive to customers", and effective at promoting conservation. As a result of the rate structure, water use in the three agencies that implemented water budget rate structures declined between 20 and 37 percent between the pre- and post-program periods (Pekelney and Chesnutt 1997).

As more utilities begin to implement water budget rates, reports and evaluations of their programs are beginning to appear in the literature. Aurora, Colorado initiated a standard "one size fits all" 3-block tiered rate system in 2002 in response to a severe drought. By 2003, Aurora determined this rate structure was inequitable for a number of customers and progressed to a water budget system for large dedicated irrigation accounts and a "fixed outdoor water budget" for all residential accounts. By 2004, individual water budgets for all customers were adopted by Aurora (Reidy 2005). The move to a water budget based rate structure in Aurora was supported by utility staff, City leadership, and citizens alike and represents a clear example of a utility striving to develop an aggressive tiered rate structure that sends a fair and equitable price signal to customers. The City of Boulder, Colorado rolled out its water budget rate structure in early 2007. This rate structure, discussed in more detail later in this report, was requested by the citizens after they were dissatisfied with the mandatory drought restrictions put in place by the City during the drought of 2002 (DeOreo, et. al. 2005).

California Pushes for Water Budgets

CUWCC BMP 5

California has taken significant steps to promote water efficiency throughout the state. The California Urban Water Conservation Council (CUWCC) was created to facilitate water efficiency statewide through partnerships and the implementation of fourteen comprehensive conservation Best Management Practices (BMPs). BMP 5, officially titled "Large Landscape Conservation Programs and Incentives", specifies the creation of evapotranspiration (ET) based water budgets for 90% of accounts with dedicated landscape irrigation meters. While these budgets are not required to be tied to water rates, developing the budgets is an important step in the overall process. A number of agencies have documented their efforts to comply with this BMP since it was modified in 1997 to include water budgets (Kah et. al 1998).

In Orange County, California, efforts have targeted personnel with direct knowledge or working contact with large landscapes to assist in the water budgeting process. Landscape maintenance contractors, homeowner association boards, property management companies, city and school district landscape maintenance staff, and others are enlisted to help develop the appropriate landscape water budget for existing landscapes which is then compared against the actual water use records provided by the retail water agency. Landscape contractors who succeed in managing water use at a site to a level at or below the specified water budget receive an "efficient irrigator" certification and can use the award in promoting their business (Berg and Ash 2002).

The City of San Diego has integrated landscape water budgets into selected customer's water bills, not as a mechanism for determining the break-point between block rates, but as an informational tool. The City uses GIS analysis tools and ET data to develop the water budgets which are mailed to customers and printed on water bills thus allowing customers to compare their actual water use to budgeted amounts (Carney et. al. 2003).

To help comply with BMP 5, the Santa Clara Valley Water District was awarded a grant to accurately measure large urban and agricultural landscape areas within Santa Clara County. Water budgets were created using multi-spectral aerial photographs and ET and information is delivered to customers via the internet (De La Piedra 2004). This type of photographic analysis is generally more applicable in arid regions where there is a visible difference between irrigated and non-irrigated areas.

California Legislation

California Assembly Bill 2717, signed into law in 2004, requested the CUWCC to convene a stakeholder task force to make recommendations for achieving greater landscape water efficiency. The recommendations from the AB 2717 task force were made public at the end of 2005 and recommend the widespread adoption of conserving rate structures and landscape water budgets (CUWCC 2005). A bill carrying forward some of the recommendations of the task force (AB 1881) was passed in California in 2006. The bill required that water budgets be included as future component of model landscape ordinances which will be updated before 2009. The text of AB 1881 is included in Appendix A of this report.

CHAPTER 3

UTILITY WATER BUDGET PROGRAMS

WHO IS USING WATER BUDGETS?

During the 1990s, water budget rate structures were only implemented by a few water agencies in California, most notably in Capistrano Valley, IRWD, and the Otay Water District. It was easy to dismiss the water budget concept during this time as another fad. After all, these three water providers are all located in southern California, were of comparable size, and were all confronted with similar drought conditions in the late 1980s and early 1990s. By 2007, water budget rate structures have diversified and expanded throughout California and across the United States to Utah, Nevada, Colorado, North Carolina, and Florida. The researchers were able to identify 25 water providers that are using some form of a water budget for billing and informational purposes.

While still primarily a California and western states phenomenon, water budgets have found support in a broad range of utilities from small systems like Marco Island, Florida to the Los Angeles Department of Water and Power (LADWP). Water budgets are no longer an exclusive "boutique" rate structure for select technologically savvy providers. These rate structures have been adopted by a broad range of providers, each adopting their own version of the water budget concept. The complete list of water budget utilities identified through this study is shown in Table 3.1.

Since drought response and water conservation are key drivers for the implementation of water budgets, perhaps it is not surprising that California and other western states have been the places where these rate structures are most popular. While these issues are of importance elsewhere in the country, water budgets have not yet been embraced as a solution in other regions.

Technical, data, and software issues do not appear to have been significant barriers for the agencies that have implemented water budgets. As can be seen in case studies presented later in this chapter, many of the implementing agencies used their existing billing system and minimal external development support to develop the software and data necessary to successfully implement water budgets. A more detailed discussion of the barriers to implementation is presented later in this report.

Table 3.1 Water providers using water budgets

Utility	Location
California	
Capistrano Valley WD	San Juan Capistrano, CA
City of Rohnert Park	Rohnert Park, CA
City of Santa Barbara	Santa Barbara, CA
Contra Costa Water District	Contra Costa, CA
Eastern MWD	Los Angeles (region), CA
EBMUD	Oakland, CA
IRWD	Irvine, CA
LADWP	Los Angeles, CA
Lake Arrowhead	Lake Arrowhead, CA
Monterey District Tariff Area	Monterey, CA
Otay Water District	San Diego (region), CA
Redwood City	Redwood City, CA
San Clemente	San Clemente, CA
Santa Rosa	Santa Rosa, CA
SDCWA	San Diego, CA
Utah, Nevada, and Colorado	
Centennial Water & Sanitation District	Highlands Ranch, CO
City of Albuquerque	Albuquerque, NM
City of Aurora	Aurora, CO
City of Boulder	Boulder, CO
City of Castle Rock	Castle Rock, CO
Las Vegas Valley Water District	Las Vegas, NV
Salt Lake City Dept. of Public Utilities	Salt Lake City, UT
Southern Nevada Water Authority	Las Vegas, NV
North Carolina and Florida	
City of Morrisville	Morrisville, NC
Marco Island	Marco Island, FL
Town of Cary	Cary, NC

CONSTRUCTION OF WATER BUDGETS

Utilities are unique and so are their rate structures, be they water budget-based or not. The formulation of water budgets and implementation within the utility context was unique for every example the researchers have identified. Only three of the identified water budget utilities developed a water budget for all customers in the service area. Most of the water budget utilities opted to use water budgets for only certain customer classes (such as single-family residential or dedicated irrigation) and some only use water budgets for informational purposes in which the budget is not linked to the rate structure. Single-family residential and dedicated irrigation accounts were the most common customer categories to be included in a water budget program. Table 3.2 summarizes the type of water budget program implemented in the identified water agencies. Details on the formulation of water budgets at all identified agencies are presented in

Table 3.3. A comparison of the rates and charges for single-family customers at selected water budget agencies is shown in Table 3.4.

Many (but not all) agencies provide a process to appeal or adjust water budgets for specific extenuating circumstances such as large families, special medical needs, unaccounted for irrigated area, or other factors.

Table 3.2 Summary of identified water budget programs in US by customer class impacted

Type of Water Budget Program	Number of Utilities Implementing	Percent of All Water Budget Utilities ID'd
Complete (all/most customer accounts)	3	11.5%
SF Residential and Irrigation Accounts	7	26.9%
SF Residential Accounts	3	11.5%
Irrigation Accounts	4	15.4%
Golf Courses	2	7.7%
Informational	5	19.2%
In process/pending	2	7.7%

Landscape Water Budgets

Landscape water budgets were the original basis for the formulation of these rate structures and are the foundation of the majority of implementations examined in this report. The idea of calculating the volume of water required to adequately irrigate a plot of land is hardly a new concept. Farmers and agronomists have been developing water budgets for agriculture for many years using evapotranspiration measurements to determine the required application depth to maximize plant growth. In the urban context, water budgets have been used for years to determine irrigation efficiency for parks and large irrigated parcels. They have been used at least since 1989 in conjunction with inclining block rate structures to provide a tailored price signal based on the water requirements of the landscape.

Landscape water budgets are typically based on two fundamental factors:

1. Landscape area
2. Water application requirement

The Centennial Water and Sanitation District in Douglas County, Colorado bases their outdoor water budgets on these two factors. Customers' budgets are calculated based on 27 inches of irrigation water applied across 45% of their measured total lot area for an entire year. The value of 27 inches was selected based on the historic evapotranspiration rate for typical landscapes in the service area. For example, a 10,000 square foot lot in Centennial would receive an annual outdoor water budget of 75.7 kgal (10,000 sf x 0.45 x 27 inches ÷ 12 inches/ft x 7.48 gal./cubic foot ÷ 1,000 gal/kgal). The water in the annual budget is distributed through the irrigation season (April – October in Colorado) based on the historic evapotranspiration curve in the area.

More complex water budget formulations are also practiced and include additional factors such as effective precipitation and different landscape water requirements. In Redwood

City, California the landscape material is taken into consideration and 100% of ET is provided for turf and 80% of ET for non-turf areas. In Santa Rosa, California, landscape water budgets are calculated by subtracting effective precipitation from the reference evapotranspiration (ET) thus creating a measurement of Net ET and then providing 100% of Net ET for high water use landscaped area and 60% for medium water use landscaped area.

Some agencies develop their outdoor water budgets using a physical measurement of the lot size or landscaped area taken from aerial photographs or GIS mapping. Other agencies use local tax assessor records of lot size to fix their customers into water budget bins. The Los Angeles Department of Water and Power (LADWP) for example has five lot size bins that customers are placed into for the creation of water budgets. LADWP also uses the customer zip code to place them into one of three temperature zones which is also used to adjust the allocation. A wide variety of formulations is possible and the method chosen by each agency is often determined by the availability of data and the desire to further customize the water budget to fit the characteristics of each landscape.

One of the criticisms frequently leveled against landscape water budgets in the residential context is that they encourage profligate use by customers with large lots by providing them with enormous water budgets based on lot size. The rate structure implemented in Boulder, Colorado tackles this issue by providing 15 gallons/square foot (gpsf) for the first 5,000 sf of landscaped area; 12 gpsf for the next 9,000 sf; and 10 gpsf for everything above 14,000 sf. Ten gpsf in Boulder is the water requirement for low water use native plants, so this approach is designed to discourage wall to wall turf on large landscapes. With all of the rate structures examined in this study, customers may use as much water as they want, but if they exceed their budget the price increases steeply.

The calculation of evapotranspiration provides another wrinkle in the formulation of landscape water budgets. Many agencies have chosen to use the historic average ET rate to calculate water budgets. This approach could potentially provide excessive budget volumes in wet years and insufficient budget volumes in dry years when the ET rate is higher than average. Using real-time ET measurements is a way around this issue and this is the approach used by the Irvine Ranch Water District (IRWD) and the Capistrano Valley Water District. In both cases, prevailing ET data is input into the billing system software and the ET for each customer's billing period is used to calculate their water budget for that billing cycle. Since it is customary to provide customers with the volume of water allocated to the upcoming billing cycle on each water bill, both IRWD and Capistrano must provide an estimate of this value and then adjust the actual budget based on prevailing conditions. Other agencies have shied away from this approach because of the perceived complexity of dealing with real-time weather data and the uncertainty this brings to the formulation of an annual landscape water budget.

Several electronic landscape water budget calculators are available on the internet for utilities to review and consumers to utilize. Typically these calculators are not associated with a water agency rate program. A listing of web-based water budget calculators is provided in Appendix B to this report.

Indoor Water Budgets

Indoor water budgets are formulated by customer class such as single-family residential, multi-family residential, commercial, etc. Developing indoor budgets for the residential sector is fairly straight forward, but the commercial, institutional, and industrial (CII) sector poses a greater challenge. As with landscape water budgets, a variety of different approaches to developing indoor budgets has been developed at different water agencies. Each customer class is addressed separately below.

Residential – Single-Family

Single-family water budgets are typically set based on the average amount used by these customers in that particular agency. Single-family indoor budgets range from 5 to 7 kgal per month depending upon the agency. Only IRWD uses a daily per capita value of 75 gcd to develop indoor water budgets for single-family customers. A family of four in IRWD would receive a monthly budget of approximately 9 kgal per month.

Residential – Multi-Family

Multi-family water budgets are developed similarly to single-family budgets, but on a per unit basis. Creating a water budget for multi-family customers requires the water utility to have valid information on the number of units at each multi-family property. This information is often available from tax assessor records and is occasionally maintained by the water provider. Multi-family indoor budgets range from 4.4 to 5 kgal per unit per month depending upon the agency. As with single-family customers, IRWD uses a daily per capita value of 75 gcd to develop indoor water budgets for multi-family customers.

Commercial, Industrial, and Institutional (CII)

The CII sector presents a greater challenge for the development of water budgets. Only a handful of agencies have included CII customers in their water budget rate structure program. Approaches to developing CII budgets usually include relying on historic usage patterns to establish a baseline budget. The problem with this approach is that it rewards customers whose use is historically inefficient with a water budget that "grandfathers" in that in-efficiency. Alternative approaches include using site-specific factors such as the number of seats in a restaurant or the number of employees in an office building to develop the water budget. This approach is impractical for many water providers who simply don't have the staff and data resources to develop water budgets in this manner.

IRWD has the most developed program of water budgets for CII customers. The IRWD CII budget program purports to be site specific and based on productivity, the number of employees, water efficiency practices at the site and other factors. In practice IRWD's CII budgets are based on historic average demands that are occasionally scaled up or down as appropriate. When a new CII customer is added to the system they are given a six month grace period from the water budget rate structure, then the next six months are used to develop a demand index (baseline budget) for the rate structure. It is anticipated that CII customers will request a budget review promptly if their usage changes because the penalty rates are significant and kick in if the water budget is exceed by more than 10%.

It is anticipated that as water budget rate structures develop over time, alternative methods for creating CII water budgets will be developed.

Table 3.3 Water budget rate structure information

Agency	Location	Type of Water Budget Program	Indoor Budget Basis	Outdoor Budget Basis	Rate Structure	Program Description
Capistrano Valley WD	San Juan Capistrano, CA	Residential and irrigation accounts.	SF/condo - 6.7 kgal/month MF - 4.5 kgal/unit/month	Outdoor allotment is based on irrigable area, ETo, crop coefficient for turf grass, management coefficient, and effective rainfall.	3 tier, increasing block rate	Two separate water budget categories – (1) for residential lots under 7,000 square feet and (2) those greater than 7,000 square feet (irrigable area > than 3,636 square feet). Indoor allotment is 9,000 ccf/month. Outdoor allotment is based on irrigable area, ETo, crop coefficient for turf grass, management coefficient, and effective rainfall.
Centennial Water & Sanitation District	Highlands Ranch, CO	Residential and irrigation accounts.	Residential - 6 kgal/month	27" x lot size.	4 tier, increasing block rate.	Four block system. For SF residential - 6,000 gal/month indoor budget. Outdoor budget = 27" of irrigation per year per lot area.
City of Albuquerque	Albuquerque, NM	Irrigation only	NA	Annual irrigation budget . Golf courses – 37"/yr. Parks – 35"/yr. Athletic fields – 45"/yr.	Surcharge for use in excess of annual budget	Applies to existing golf courses, city owned parks, and city owned athletic fields. Two block structure. Excess of water budget = 50% increase in commodity rate. Planting restrictions, but not water budgets, apply to new development.
City of Aurora**	Aurora, CO	Complete - all customers	SF - 7 kgal/month Others - 70% of historic avg.	SF Outdoor = ET x 3,000 sf irrigated area. Appeals process available for larger lots		3 block structure. Indoor budget based on most recent winter quarter average. Outdoor based on 70% of 00-01 consumption. New accounts receive 7,000 gal/month indoor (3.2 people x 70 gcd). Outdoor = ET x 3,000 sf irrigated area.
City of Boulder	Boulder, CO	Complete - all customers	SF - 7 kgal/month MF - 5 kgal/month ICI - historic avg.	Lot size. 15 gal/sf for first 5,000 sf landscape area. 12 gal/sf for next 9,000 sf. 10 gal./sf for everything above 14,000 sf.	5 tier, increasing block rate	Full fledged water budget rate structure starting in 2007
City of Castle Rock	Castle Rock, CO	In process of implementing water budget rates.			In process of implementing	The City of Castle Rock is considering the use of water budgets as part of its Water Resources Strategic Master Plan, November, 2005 in an effort to decrease water consumption by 18% over the next 25 years.
City of Morrisville	Morrisville, NC	SF residential and irrigation			4 tier, increasing block rate	Water budget system that is the same as the one implemented in the Town of Cary; the only difference is the price of the lowest tier – up to 5,000 gallons. The cost for this tier is $3.75/Kgal; in the Town of Cary the price is $3.28/Kgal.

(continued)

Table 3.3 (Continued)

Agency	Location	Type of Water Budget Program	Indoor Budget Basis	Outdoor Budget Basis	Rate Structure	Program Description
City of Rohnert Park	Rohnert Park, CA	Informational - irrigation only	NA	Based on landscape area, ET	NA	Landscape water use report provided to dedicated irrigation meter sites each billing period.
City of Santa Barbara	Santa Barbara, CA	Irrigation only	NA	Annual allotment based on irrigated area.	3 tier, increasing block rate	Irrigation only accounts receive budget based on acreage. Budget fits into two tier system. Rates differ based on customer class.
Contra Costa Water District	Contra Costa, CA	Informational - irrigation only	NA	Based on landscape area, ET	NA	
Cottonwood Water and Sanitation District	Douglas County, CO	Residential.	SF - 6 kgal/month MF - 4.4 kgal/month	30" x 2,900 sf.	3 tier, increasing block rate.	Three tier rate structure with single-size water budget based on 2,900 SF of irrigable area. Appeals process available for large lots.
Cucamonga Valley Water District	San Diego, CA	Residential - pilot				Pilot program to create water budgets for 300 single-family customers.
Eastern MWD	Los Angeles, CA	In process of implementing water budget rates.			In process of implementing water budget rates.	
EBMUD (current)	Oakland, CA	Informational - irrigation only	NA		NA	Landscape water budget program since 2002. Informational only. Does not impact the water bill.
EBMUD (historic 1988-92)*	Oakland, CA	Irrigation only – drought response program	NA	Based on irrigable area, 80% of ET, goal was to achieve a 50% cutback from previous/baseline year	5 tier, increasing block rate	Customer submitted landscape area and the utility verified. 1,200 landscape accounts including public agencies and golf courses participated in program. Efficiency standard was based on 80% of ET. Rates were used to penalize. Rates were based on % cutback (50%). Exceeding target became expensive. Because of billing system constraints the system required manual data entry which was time consuming.
IRWD	Irvine, CA	Complete - all customers	SF/MF - # of res. x 75 gpd ICI - site specific based on historic use, productivity,	ET x Kc x 1.25 x landscape area	5 tier, increasing block rate – Res. & Lscape. 4 tier,	Key example of rate structure implementation. Five tier block rate structure. Water budget based on standard indoor amount and outdoor on landscape area.

(continued)

Table 3.3 (Continued)

Agency	Location	Type of Water Budget Program	Indoor Budget Basis	Outdoor Budget Basis	Rate Structure	Program Description
			employees, etc.		increasing block rate - ICI	
LADWP	Los Angeles, CA	SF residential		Based on lot size and temperature zone. Five bins.	2 tier, seasonal rate	Two tiers: Tier 1 is based on lot size (5 blocks), temperature zone (3 zones) which is determined by zip code, and a 7 block household adjustment for households with 7 or more persons. Tier 1 rates are in effect from November 1 through May 31.
Las Vegas Valley Water District	Las Vegas, NV	Golf courses	NA			Currently applies only to golf courses. http://www.lvvwd.com/html/ws_drought_restrictions_golf.html
Marco Island	Marco Island, FL	Residential and irrigation accounts.	Residential - 6 kgal/month	Based on lot size. Five bins.	5 tier, increasing block rate	Home to 15,000 permanent residents and another 35,000 seasonal residents, Marco Island is located in southwestern Florida. On June 6, 2005 the city council adopted a 3 block rate structure with an allocation of 6,000 gallons/month for indoor use and a 5-tier system of lot size.
Monterey District Tariff Area	Monterey, CA	Residential and irrigation accounts.	Based on number of residents and large animals.	Based on lot size and temperature zone. Eight bins.	5 tier, increasing block rate	The water budget (also known as the ECU or equivalent consumption unit) for these areas is based on the number of people in the household, the size of the lot, an allotment for large animals, and a seasonal adjustment. The formula for determining the ECU can be found at http://www.calamwater.com/awpr1/caaw/pdf/Rates_MO-1_combined.pdf
Otay Water District	San Diego, CA	Informational - irrigation only	NA	Monthly information on application rate.	NA	Had water budget program until 2000. Currently have voluntary landscape water budget program.
Redwood City	Redwood City, CA	In-process. Budgets are being developed for residential, commercial, and irrigation accounts	SF/MF – per capita allocation CII – case by case	Based on surveyed irrigated area, 100% of ET for turf, and 80% of ET for non-turf.	To be decided. Will be 4 or 5 tier structure.	Transitioning to water budget rate structure as part of conservation and drought response programs. Water budgets designed to provide customers with adequate water for reasonable use. Charges for water use within the budgets are strictly based on the cost of service, but use for water above the budgets is charged at much higher rates (both marginal costs for new firm supplies or penalty rates) with the intention of discouraging wasteful use.
San Clemente	San Clemente,	Residential and irrigation	SF/condo - 6.7 kgal/month MF -	Outdoor allotment is based on irrigable area, ETo, crop	Seasonal, three tier increasing	Single family residents are billed monthly using a 3-tier system based on lot size. Lot size is considered in

(continued)

Table 3.3 (Continued)

Agency	Location	Type of Water Budget Program	Indoor Budget Basis	Outdoor Budget Basis	Rate Structure	Program Description
	CA	accounts.	4.5 kgal/unit/month	coefficient for turf grass, management coefficient, and effective rainfall.	block rate.	customer classification with price break points at 7,000 square feet, 9,000 square feet and 14,000 square feet. Allotments vary with summer and winter use in each of the 3 tiers for the different lot sizes.
Santa Rosa	Santa Rosa, CA	Irrigation only	NA	{(ET- Effective precip.) x (Landscape area high use + (Landscape area medium use x 0.6)) x conversion factor}	3 tier, increasing block rate.	
SDCWA	San Diego, CA	Informational	NA		NA	Businesses participating in irrigation controller incentive programs will be enrolled in a free water budget program that informs customers of their water use in relation to a water budget equal to 80 percent of reference evapotranspiration for the landscaped area. Sites may enroll in the free water budget program even if they do not participate in the incentive programs.
Salt Lake City Dept. of Public Utilities	Salt Lake City, UT	Irrigation only	NA	Water use target based on irrigated area, ET, landscape coefficients, and demand levels.	2 tier, increasing block rate	Water used within the target is billed per unit in tier 2; water consumption exceeding the target is billed per unit in tier 3. During declared water shortages, irrigation-only accounts may be called upon to make "measurable reductions".
Southern Nevada Water Authority	Las Vegas, NV	Golf courses	NA	Between 75" and 78"/year depending on drought alert status.	4 tier increasing block rate	Water budgets are calculated based on the number of acres irrigated. This includes all lakes and ponds within the course and those used all or in part as an irrigation reservoir. Budgeted acre-feet include potable, raw, reclaimed and recycled water.
Town of Cary	Cary, NC	SF residential and irrigation	SF - 5 kgal/month	Based on landscape area, ET, rainfall, efficiency factor	4 tier, increasing block rate	Four tier rate system for single family residents – the lowest block is for water use up to 5,000 gallons – with charges increasing fairly gradually for the first 3 tiers up to 23,000 gallons. The price for tier 4 more than doubles in price. The price for the tiers is based on an average lot size of 0.25 acres and an annual irrigation allotment of 15,000 gallons.

*First identified implementation of a water budget rate structures.
**Rate structure modifications for 2007, not reflected in this table.

Table 3.4 Comparison of rates and charges at selected utilities implementing water budgets

Utility	Fixed Charges (per month)	Block 1 ($/kgal)	Block 2 ($/kgal)	Block 3 ($/kgal)	Block 4 ($/kgal)	Block 5 ($/kgal)	Sewer Rate if applicable
City of Boulder, CO	$ 8.55	$ 1.88	$ 2.50	$ 5.00	$ 7.50	$ 12.50	$3.50/kgal indoor
Centennial Water and Sanitation District	$ 12.50	$ 2.30	$ 3.25	$ 5.20	$ 7.80		$10.25/month fixed plus $2.29 per 1,000 gallons
Irvine Ranch Water District, CA	$ 6.75	$ 1.00	$ 1.22	$ 2.42	$ 4.87	$ 9.73	Residential: $9.80 - $13.05/month depending upon usage CII: $15 for first 7.5 kgal and $1.22 for each additional kgal.
Los Angeles DWP (seasonal rates)		$2.85 - $2.99	$3.40 - $4.27				$3.81/kgal indoor
Capistrano Valley Water District, CA	$ 11.42	$ 3.03	$ 3.89	$ 6.22			15.42/month
Town of Cary, NC	$ 2.76	$ 3.28	$ 3.75	$ 5.33	$ 10.83		$4.41/kgal
San Clemente, CA	$ 7.26	$ 1.63	$ 2.44	$ 3.66			$16.27/month (fixed)
Marco Island, FL	$ 24.11	$ 3.02	$ 4.53	$ 6.04			$19.67/month fixed plus $3.91/kgal (6 kgal max)
Monterey Area Tariff District, CA	$ 6.91	$ 2.19	$ 4.38	$ 8.76	$ 17.53		

CHAPTER 4

WATER BUDGET PROGRAM CASE STUDIES

This chapter presents illustrative case studies that describe the water budget rate structure implementation in a number of cities. These case studies were selected to provide detailed information on how water budgets have been applied in a broad diversity of water agencies. The researchers have also included one case study, Otay, which reviews a water budget program that was rescinded and then re-implemented in a substantially revised form. There are important lessons to be learned from the Otay experience as well as from the on-going water budget programs.

CASE STUDY: CITY OF SAN JUAN CAPISTRANO

One of the most mature water budget rate structure programs in the country has been implemented in the City of San Juan Capistrano since 1991. While perhaps not as well known as the water budget program in IRWD, the Capistrano rates structure has served as a model for numerous agencies across California and the western U.S. seeking to implement comparable programs. The San Juan Capistrano rate structure is a success story in that it achieved significant water savings (18.6 – 22% according to Pekelney and Chesnutt, 1997), maintained required utility revenues, and has been popular with residents for 15 years.

San Juan Capistrano (SJC) is a small city of nearly 40,000 people located in south Orange County, California north of San Diego and south of Los Angeles. SJC's service area is made up of primarily residential development and open space with small areas dedicated to commercial, agricultural and public/institutional use. SJC is the site of the oldest settlement in Orange County, and is proud of its strong roots in its Spanish, Mexican and early California heritage.

In 2005, SJC delivered 11,080 acre-feet (3612 million gallons) of metered water to 10,870 customer accounts. The City recently constructed San Juan Groundwater Recovery Project (GWRP), also referred as San Juan Basin Desalter Project. The GWRP is a reverse osmosis filtration/treatment facility that will produce up to 4,800 acre-feet of potable water per year. This represents more than half of the City's annual potable water requirement as well as a significant portion of its emergency water storage needs. Until construction of GWRP, approximately 92 percent of the City's potable water supply was purchased from the Metropolitan Water District of Southern California (MWD) through MWD Orange County. The remaining 8 percent was derived from local groundwater sources. Construction of the GWRP has significantly changed these percentages.

Implementation Process

SJC's water budget rate structure was developed in response to the California drought of the late 1980s and early 1990s. In 1990 MWD required all member agencies to reduce demand by 30% or face surcharges and possibly curtailment of supply. Under the leadership of SJC's General Manager, Ray Auerbach (also an elected board member at IRWD) the water budget rate structure concept was developed and implemented as a way to fairly achieve the demand reductions required by MWD and respond to the drought.

Water conservation was in the news daily at the time of the drought and SJC's Water Advisory Commissioners were supportive of the staff proposal for a water budget rate structure. The public showed strong interest in the proposal and attendance was high at Water Advisory Commissioners meetings. Final adoption from the city council (acting as the Water District Board of Directors) occurred with unanimous support.

Implementation of the rate structure and billing system took a concerted effort on behalf of the SJC staff. General Manager Ray Auerbach noted that it wasn't the hardware or software that made the accomplishment possible; it was the "peopleware" – meaning the SJC staff. Only limited programming additions to SJC's billing system were required and the water bill itself was re-designed as part of the implementation process.

The water budget rate structure implemented in SJC (and still in place) is a three tiered increasing block rate structure with water budgets for all residential customers and dedicated landscape and agricultural meters/accounts. The allocation for residential accounts includes an indoor component plus an outdoor component. The outdoor allocation changes each month depending upon prevailing weather conditions and is based on irrigable area, ETo (reference evapotranspiration), a crop coefficient for turf grass, a management coefficient, and effective rainfall. Single-family residences were initially divided into lots less than 7,000 square feet and lots greater than 7,000 square feet. Lots fewer than 7,000 square feet are all given the same allotment based on 3,636 square feet of irrigable area. Lots greater than 7,000 square feet are given an individualized water budget based on their landscape area. The single-family indoor allotment is 9 ccf (6.7 kgal) per month and the multi-family indoor budget is 6 ccf (4.5 kgal) per month.

Table 4.1 shows the current block rate structure for residential (SF and MF) and dedicated landscape accounts in San Juan Capistrano. Use up to the budget/allocation is charged at the tier 1 rate. Use up to 200% of the budget is billed at the tier 2 rate. Use greater than 200% of the budget is charged at the tier 3 rate.

Table 4.1 San Juan Capistrano water budget block rate charges (effective July 1, 2006)

Tier	Cost per CCF of water
Tier I: Use up to 100% of the allocation	$2.27 per ccf
Tier II: Use up to 200% of the allocation	$2.91 per ccf
Tier III: Use greater than 200% of the allocation	$4.85 per ccf

SJC held numerous public meetings and made frequent use of the utility's quarterly newsletter, along with bill stuffers and newspaper articles to publicize the new rate structure. Trained customer service staff members were in place to handle questions and complaints that came in. SJC did receive a small number of customer contacts after implementation of the rate structure and most of the customers were seeking lot size adjustments.

Francie Kennedy, Water Conservation Coordinator for San Juan Capistrano, has managed the program almost since its inception, which gives her a thorough knowledge of its details. Kennedy has worked directly with customers impacted by the rate structure and has had significant input into changes and adjustments made by the utility over the past 15 years.

Kennedy explained that SJC began adjusting their rate structure as soon as it was implemented. Adjustments were made for large lot customers and for multi-family properties with common area irrigation. In 1992, the drought ended and MWD relaxed demand restrictions. SJC responded to these relaxations by increasing allocations to its residential customers, and switching to real-time weather adjustments for irrigation allocations. SJC does not have a formal appeals process for customers who are dissatisfied with their water allotment. Instead, Kennedy handles these issues on a case by case basis in person or over the phone. Most of the initial requests related to the lot size and occasionally Kennedy made site visits to re-measure areas. SJC also makes billing adjustments if a leak is found and the customer repairs it.

Billing System and Software

Implementation of the SJC water budget rate structure did not require new billing software. Instead, programming to the existing system was accomplished and new billing forms were created at an estimated cost of $5,000. The billing software has subsequently been upgraded due to changes in hardware and operating systems.

At the outset of program implementation, SJC purchased a database of lot sizes for approximately $1,000. These data were used to develop the first outdoor allocations. A part-time college student employee was hired for several months to assist with area measurements and verification. Since the initial implementation the SJC water budget billing system has not required any additional staff to operate.

A sample water bill from SJC is provided in Figure 4.1. Allocation information is shown in the upper left hand box. The current month's allocation is shown along with an estimate of the next month's allocation. It is only an estimate because SJC adjusts the allocation each month to match local climate conditions as explained above. The bill also provides information on the lot size and building footprint. Directly below the "Allocation Info" box on the bill is a use history that provides approximately 1 year of billing history for the customer. This is also the section of the bill that shows the volume of water consumed in each of SJC's three rate tiers. Please note that this sample bill is a "final bill" for the customer and as such had extra final billing information not found on a regular monthly bill.

City of San Juan Capistrano
Water Services Department
32400 Paseo Adelanto
San Juan Capistrano, CA 92675
949/493-1515

KEEP THIS PORTION FOR YOUR RECORDS

STATEMENT

CUSTOMER NO: 021749

Account No: 0035071

BILLING DATE: 4/27/2005 SERVICE PERIOD 4/13/2005 TO 4/22/2005

07 WCB 1

SERVICE ADDRESS: [redacted]

METER INFORMATION #	Current Reading	Previous Reading	Usage in CCF (1 ccf =748 Gallons)	# of Days	Usage in Gallons
47884369	4092	4071	21	9	15708

Final Billing

Allocation Info:

This Month's Allocation (ccf): 33

Next Month's Est. Allocation (ccf): 19

Lot Size (sq ft): 15,507

House Size (sq ft- ground floor): 4,017

Use History

Billed	Allocation	Use-ccf
Apr 22, 2005	33	21
Apr 13, 2005	33	40
Mar 16, 2005	18	33
Feb 10, 2005	19	15
Jan 12, 2005	16	18
Dec 13, 2004	18	24
Nov 10, 2004	19	19
Oct 11, 2004	32	52
Sep 10, 2004	39	66
Aug 10, 2004	44	69
Jul 12, 2004	39	47

Tier 1 Usage	10	ccf @ $1.97	=	$19.70
Tier 2 Usage	10	ccf @ $2.52	=	$25.20
Tier 3 Usage	1	ccf @ $4.20	=	$4.20

Previous Balance	$107.95	
Past Due Total - Pay Immediately to Avoid a Late Charge		**$107.95**
Water Use Charges	$49.10	
Water Service Charge	$2.97	
Sewer Use Charges	$4.62	
	Current Charges	**$56.69**

Current Charges, Due By: 5/12/2005

TOTAL DUE > **$ 164.64**

(SEE REVERSE SIDE FOR IMPORTANT INFORMATION, MAP AND PHONE NUMBERS)

PLEASE RETURN THIS PORTION WITH YOUR PAYMENT

BILLING DATE: 4/27/2005

Account Number : [redacted]

CUSTOMER NUMBER: [redacted]

Service Address: [redacted]

Note Area

PLEASE ENTER AMOUNT PAID >

City of San Juan Capistrano
Water Services Department
32400 Paseo Adelanto
San Juan Capistrano, CA 92675
949/493-1515

PREVIOUS BALANCE > $107.95

CURRENT CHARGES > $ 56.69

PLEASE PAY THIS AMOUNT > $ 164.64

CHECKS MAY BE MADE PAYABLE TO "CITY OF SJC"

[redacted]

SAN JUAN CAPISTRANO CA 92675

021749003507100001646405122005

9/29 > Collections

00634020-35

Figure 4.1 SJC sample water bill

15 Years of Operational Experience

The SJC water budget rate structure has been in place for 15 years and continues to be an effective billing methodology and water conservation tool for the agency.

"Customers who have questions usually perceive the rate structure as fair after it is explained," Kennedy said. "Most customers are able to reduce their use when they get into tier 2 or 3 and many leaks have been found and repaired as a result of our rate structure. Of course, a few customers can afford to pay anything, don't care about the water bill, and just use whatever they want."

"At the beginning of the program it was a real eye-opener for customers to learn how they used water vs. what they should have been using," Kennedy explained. Kennedy notes that the program has been effective at stimulating customers to take water efficiency seriously and do things like repair leaks, replace old toilets, and improve irrigation practices.

"The rate structure communicates the value of the resource," Kennedy said. "It discourages waste, reduces runoff, and has reduced hardscape[9] damage due to over-irrigation." From the utility revenue perspective, Kennedy noted that the rate structure has made it easier to develop accurate sales and revenue projections which in turn have improved the budgeting process.

Recommendations for Other Agencies

Kennedy has a few recommendations to any agencies considering implementing a water budget rate structure. These include:

- Research data needs (landscape areas, weather, etc.) and get these in place ahead of time.
- Get the facts right. In particular defensible weather data, appropriate horticultural science, accurate measures of landscape areas, etc.
- Be sure the billing system can handle the inclusion of a water budget for each customer.
- Educate customer service staff to handle customer questions and complaints.
- Educate customers to the maximum extent possible about the rate structure, the way they're billed, the scientific foundation of the allocation, and other related issues. Provide as much advance notification and information as possible.
- Make sure everyone on the utility staff is behind the effort from general manager, to department heads, to programmers, customer service representatives, and conservation specialists.

CASE STUDY: IRVINE RANCH WATER DISTRICT

The Irvine Ranch Water District (IRWD) located in Orange County California, south of Los Angeles serves approximately 322,000 people in the city of Irvine and portions of Tustin, Newport Beach, Costa Mesa, Orange and Lake Forest. Covering approximately 179 square miles from the Pacific coast to the foothills, IRWD has employed one of the best-known examples of a water budget rate structure since 1991. The IRWD service area is in a desirable geographical and economic location and has high growth potential. Water conservation is important to IRWD because the potential avoided costs are high. IRWD's water efficiency goal is, "To be a leader in the innovation and implementation of conservation measures promoting the most efficient use of water both on a per capita and per acre basis." After 15 years and a number of adaptations and adjustments, the water budget rate structure remains a cornerstone of IRWD's strategic water management efforts and it enjoys the continued enthusiastic support of

[9] Hardscape includes impermeable surfaces in a landscape such as pavements, walkways, flagstones, and other non-plant features.

the Board of Directors, agency staff, and customers. IRWD views water conservation as a way of life and the water budget-based rate structure is considered an important on-going water management tool.

1991 – Water Budgets as Drought Response

The drought of the late 1980s and early 1990s forced the Metropolitan Water District (MWD) of Southern California, IRWD's wholesale supplier, to implement penalties for agencies that exceeded a set allotment. In response, the IRWD board instructed the staff to develop a water budget-based rate structure designed to penalize customers who use substantially more water than is required to maintain their landscapes. The IRWD staff took up the challenge and created a five tier increasing block rate structure where the block break points are determined by individual water budgets created for each customer class or subgroup (i.e. single-family detached). The board approved the plan and in March 1991 IRWD moved from a flat rate system to a two tiered increasing block rate structure as an interim measure. In July 1991 a five tier block rate water budget structure was implemented. Although the drought ended in early 1992 with the 13 inches of rain affectionately known as the "March miracle", IRWD retained the allocation based rate structure as a permanent water management tool.

The rapid implementation of the data-intensive and fairly complex IRWD rate structure was tackled in-house by IRWD staff without paying for outside consultants or programmers. "We had three people in a room working intensively for several months," said Dave Ferguson, Director of Special Projects for IRWD and one of the original architects of the rate structure. IRWD staff made the necessary programming changes to the IRWD billing system software and designed a bill format that could accommodate the elements of the new rate structure and provide useful information to customers.

"Having detailed information about customers and a sophisticated billing system improve your chances of success," Ferguson noted. "IRWD's initial ascending block rate structure was fairly primitive. There have been a number of improvements and refinements to the system implemented over the years that have resulted in a more sophisticated, science-based, and equitable system."

Water budgets were initially constructed for single-family customers using a fixed indoor volume (75 gpd x 4 people per household ~ 110 kgal/year or 9 kgal/month) and a default lot size of 1,300 square feet. Understanding that this could be problematic for customers with large families, large landscapes or special medical needs, IRWD implemented an appeals process that allowed customers to apply for a budget variance and to substitute data such as their actual lot size for the default value. According to Fiona Sanchez, IRWD Conservation Supervisor, over the 15 year life of the rate structure approximately 40% of residents have requested that their actual irrigated area be substituted for the default value. Table 4.2 shows the current allocations IRWD uses and the definition of terms are shown below the table.

Variances to the allocation formulas above are available for customers with large families, large landscapes, or special medical needs. Additionally variances are available for licensed care facilities, fire control zones, qualified CII customers, and for new account establishment.

IRWD implemented its water budget rate structure using existing resources and without adding any staff. Eventually the water conservation program was "beefed up" because it was understood that an outreach effort would make the rate structure an even more effective efficiency tool. Because of the drought conditions in California, water conservation was a high profile issue at the time the structure was put into place. Customers were informed of the new rate structure via bill stuffers, a monthly newsletter, and a mailer that was sent to each customer. Great effort was made to ensure the public knew why IRWD was implementing a water allocation budget, how it was structured, what it was intended to do, and how and when it would be implemented.

Table 4.2 IRWD Base allocations (water budgets) for commodity charges

Account Type	Base Allocation No. of Residents	Landscape Area (LA)	Base Allocation Indoor	Base Allocation Outdoor	Total Allocation
Residential Detached	4	1,300 sq. ft.	# Residents x 75 gpd	ET x Kc x 1.25 x LA	(Indoor + Outdoor) x # days in bill service period
Residential Attached*	3	435 sq. ft.	# Residents x 75 gpd	ET x Kc x 1.25 x LA	(Indoor + Outdoor) x # days in bill service period
Apartments*	2	N/A	# Residents x 75 gpd		Indoor x # days in bill service period
Irrigation		Site specific based on irrigated acreage	N/A	ET x Kc x 1.25 x LA	Outdoor x # days in bill service period
Commercial, Industrial, Institutional			Site specific, based on productivity, employees, water use efficiency practices etc.	Site specific, based on irrigation needs.	Site specific, adjusted for # days in bill service period

*For master-metered apartments and condominiums, the base allocation is multiplied by the number of dwelling units.
gpd = gallons per day
ET = evapotranspiration, taken from IRWD weather stations located in coastal central or foothill zones.
Kc = crop coefficient, the relative amount of water cool season turf needs at various times of the year
1.25 = extra water to make up for irrigation system inefficiencies (estimated at 80%)
LA = landscape area. IRWD assumes that 100% of the landscape is cool-season turf-grass

IRWD's current rate structure is essentially the same 5-tier increasing block rate system first implemented in 1991. A number of changes and adjustments have been made (particularly to the rates) over the years, but the fundamental concepts have been in place for 15 years. IRWD employs a different set of block rate commodity charges for each customer class. CII customers have a 4-tier rate rather than the 5-tier used for all residential and landscape accounts. Table 4.3 shows the rate for residential detached dwellings (current as of 7/24/2006).

Table 4.3 IRWD commodity charges for residential detached dwellings

Tier	Rate/CCF	Indoor Usage In CCF	Total Allocation Percentage of Usage*
Low Volume	$0.75	0-5	0-40%
Base Rate	$0.91	6-12	41-100%
Inefficient	$1.82	13-18	101-150%
Excessive	$3.64	19-24	151-200%
Wasteful	$7.28	25+	201+%

*The percentage of usage is based on the total allocation which includes indoor and outdoor.

IRWD's rate structure is designed to reward customers who use water within their monthly allocation with lower rates. The "low volume" tier rate is priced less than IRWD's melded cost of water and the loss is offset by penalty revenues. IRWD's base rate is among the lowest commodity charges in Orange County, California. The "base rate" tier is used as a multiplier for subsequent tiers. The "inefficient" tier rate is 200% of the base rate; the "excessive" tier rate is 400% of the base rate; and the "wasteful" tier rate is 800% of the base rate. The "wasteful" tier was originally named the "abusive" tier in the first version of IRWD's rate structure. In the late 1990s the IRWD Board elected to rename the "abusive" tier and it is now know as the "wasteful" tier.

The rate structure is somewhat different for commercial, industrial, and institutional customers. As shown in Table 4.2, allocations for CII customers are, "Site specific, based on productivity, employees, water use efficiency practices etc." In practice they are based on historic average demands that are scaled up or down as appropriate. A four tier rate structure is used for this customer class because IRWD recognizes this budgeting method is less precise than what is used for residential and irrigation accounts. When a new CII customer is added to the system they are given a six month grace period from the water budget rate structure, then the next six months are used to develop a demand index (baseline budget) for the rate structure. Fiona Sanchez noted that, "The customer will request a review if their usage changes" because the penalty rates are significant and kick in if the water budget is exceeded by more than 10%. The IRWD rate structure for CII customers is shown in Table 4.4. CII customers who use more than 120% of their allocation in any given month reach the "wasteful" tier and pay $7.28 per ccf.

Table 4.4 IRWD commodity charges for CII customers

Tier	**Rate/ccf**	**Base Index Percent of Usage**
Base Rate	$0.91	0-100%
Inefficient	$1.82	101-110%
Excessive	$3.64	111-120%
Wasteful	$7.28	121+%

IRWD has over 5,000 dedicated landscape irrigation meters. For these customers IRWD was able to establish well defined, scientifically-based water-budget allocations. Fiona Sanchez explained that since the dedicated landscape allocations are based on scientific research the resulting landscape allocations are quite precise and the rate tiers are tighter than they are for residential accounts. The current allocations and rate for irrigation accounts is shown in Table 4.5. The block breakpoint percentages are an amalgamation of the residential and CII structures. Customers who exceed their allocation by more than 20% during a billing period land in the highest "wasteful" tier. IRWD feel comfortable with this system because the irrigation landscape water budgets are well defined and carefully constructed.

Another feature of the IRWD rate system is the intentional separation of fixed and variable costs. In an effort to smooth the overall revenue stream, IRWD decided to cover fixed costs through the fixed monthly charges to customers that appear on the water bill separate from the commodity charges discussed above. IRWD's variable costs for things such as water (purchased from MWD), chemicals, energy, pumping, and labor directly related to volume.

Fixed costs typically include those non-volumetric items not included above. A sample water bill from IRWD is shown in Figure 4.2.

Table 4.5 IRWD commodity charges for landscape irrigation customers (potable*)

Tier	Rate/ccf	Base Index Percent of Usage
Low Volume	$0.75	0-40%
Base Rate	$0.91	41-100%
Inefficient	$1.82	101-110%
Excessive	$3.64	111-120%
Wasteful	$7.28	121+%

*Reclaimed irrigation rates are 10% lower than the potable rates shown here.

Response to IRWD Rate Structure

The IRWD rate structure is considered a great success by the customers, staff, and board of directors. Both Ferguson and Sanchez said that given the evolution and advancement of the rate structure over time there is nothing that they would change about the system at the moment. They noted that the residential customers elect the IRWD board, offering a regular opportunity for a referendum on the water budget rate structure. In the years since the rate structure was implemented, board members have been consistently re-elected.

"Customers perceive the rate structure as fair and equitable. They like it," said Dave Ferguson. "It's not hard to sell water conservation here." He added that surveys of customers conducted by IRWD regularly rate satisfaction with the agency "in the high 90 percent range."

IRWD has used revenues generated from the penalty tiers of the rate structure ("inefficient", "excessive", and "wasteful" tiers) to fund water conservation, water recycling, and other efficiency related projects. These revenues have also been used to offset the low volume tier which provides the first portion of a customer's allocation at a subsidized rate.

The IRWD water budget rate structure has been a tremendously effective water conservation measure and tool. Analysis by IRWD staff concludes that the program has increased outdoor efficiency by 60 percent. In 1992 shortly after the rate structure was implemented IRWD customers irrigated approximately 3,600 acres in the service area and applied roughly 4.4 acre-feet of water per acre (52.8 inches of water applied). The typical ET requirement for turf grass in the area is 48 inches per year. In 2005 IRWD customers irrigated 11,768 acres of landscape and applied only 1.7 acre-feet of water per acre (20.4 inches). This represents a 61% reduction in the outdoor application rate since the introduction of the water budget program. An independent evaluation in 1997 found a 37% decline in water applied to landscapes as a result of IRWD's rate structure (Pekelney and Chesnutt, 1997).

The time period since the rate structure was implemented has seen the rise in popularity of Xeriscape and water-wise landscaping techniques as well as a number of other changes that could also impact outdoor irrigation. The water budget rate structure encourages customers to implement these techniques. Although a number of factors may contribute to the overall changes in usage patterns, the substantial reduction can't be ignored or passed off as insignificant. There is strong evidence that the rate structure has played a leading role in the reduction of outdoor water waste.

IRVINE RANCH WATER DISTRICT

15600 SAND CANYON AVE IRVINE, CA 92618-3102
MAILING: P.O. BOX 57500 IRVINE, CA 92619-7500

PLEASE RETURN THIS PORTION
WITH YOUR PAYMENT

Web Site: www.irwd.com
BILLING INQUIRIES (949) 453-5300
8:00 A.M. TO 5:00 P.M. Monday - Friday

[redacted]

[redacted] $521.27

IRVINE RANCH WATER DISTRICT

ACCOUNT NAME [redacted]
SERVICE ADDRESS
BILLING PERIOD FROM 12/27/06 TO 1/26/07

IR ACCOUNT NUMBER [redacted]
METER NUMBER
BILL TYPE REGULAR BILL

Previous Read	Current Read	Usage (in CCF)	Usage in Gallons 748 Gallons = 1 CCF
44605	44989	384	287232

USAGE - LOW VOLUME	129 @ .750	$96.75	
USAGE - CONSERVATION BASE RATE	193 @ .910	$175.63	
USAGE - INEFFICIENT	32 @ 1.820	$58.24	
USAGE - EXCESSIVE	30 @ 3.640	$109.20	$439.82
WATER SERVICE CHARGE			$81.45

YOUR ALLOCATION FOR THIS BILL IS 322 CCF
BILL CALCULATION BASED ON 4.34 ACRES

WATER USAGE REVENUE ALLOCABLE TO
URBAN WATER RUNOFF TREATMENT $439.82 @ 2% $8.80

TO AVOID LATE CHARGE PAY BEFORE 2/19/07 $521.27

WE AT IRWD WISH YOU A
SUCCESSFUL AND PROSPEROUS 2007.

COMPARE YOUR AVERAGE DAILY USAGE IN CCF WITH LAST YEAR

FROM	TO	DAYS	USAGE BILLED	DAILY AVERAGE
12/27/06	1/26/07	30	384	12.80
12/28/05	1/26/06	29	207	7.13

WATER USE PROFILE

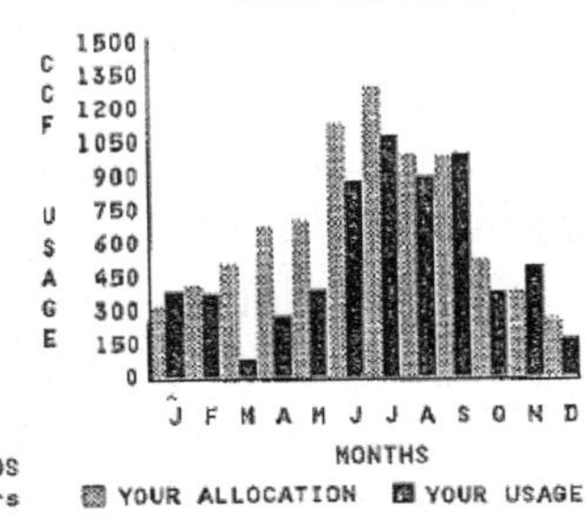

ACCT - [redacted] PLEASE RETAIN THIS PORTION FOR YOUR RECORDS
Please See Reverse Side For Important Information And Phone Numbers

Figure 4.2 Sample water bill from IRWD

Lessons Learned

The water budget rate structure remains a cornerstone of IRWD's strategic water management efforts and it enjoys the continued enthusiastic support of the Board of Directors, agency staff, and customers. When asked what advice he might offer to other agencies considering a similar rate structure, Dave Ferguson noted three things:

1. It helps to implement (or start implementation) at the time of a drought because of the focus on water efficiency and savings that occurs.
2. The utility's governing body must be committed to a water conservation management philosophy and be willing to make the change from the status quo to a new and different billing system.
3. Equity between customer groups is critical. The proposed rate structure must be viewed by the public as fair to all, and no one customer group should be required to do more or less than any other customer group.

Ferguson also said that monthly meter readings and billing rather than bi-monthly is preferred for this type of rate structure. Monthly billing provides quicker information and feedback on water use to the customers and enables more rapid correction of leaks and inefficient practices that can lead to high bills under the water budget system.

Although drought has not revisited the IRWD service area since the water budget rate structure was implemented, the agency is prepared to take advantage of the features of the system to manage any future droughts. "We can modify the allocation and the rate or both," explained Fiona Sanchez. In the simplest scenario the tiered rates stay the same, but allocations (budgets) are reduced to match IRWD's drought response goal. Demand can be easily monitored to make sure targets are met.

CASE STUDY: OTAY WATER DISTRICT

The Otay Water District (Otay) covers 129 square miles in southeastern San Diego County, California and serves approximately 186,000 people. Otay is an example of a water agency that implemented a water budget rate structure and then several years later decided to move away from water budgets. In 2006, Otay has a voluntary, informational water budget program, but does not utilize the integrated water budget rate structure it employed for most of the 1990s. Otay represents an important example of how a water budget rate structure has been implemented and the forces of change that can move an agency in a different direction.

1992 – Otay Adopts Water Budgets

For many water agencies in southern California, the late 1980s and early 1990s were a period of drought that required significant action to reduce demand and provide adequate supply. In response to the drought, Otay convened a task force of utility staff and landscape professionals from the area including the City of Chula Vista[10]. Jan Tubiolo, an expert in horticulture and landscape irrigation, was the water conservation coordinator for Otay in 1991. Tubiolo served

[10] The City of Chula Vista is served by the Otay Water District.

on the task force and was instrumental in developing and implementing Otay's water budget rate structure.

At the time, 17% of Otay's customer accounts were for dedicated landscape irrigation. Furthermore, it was estimated that these customers accounted for more than 40% of the overuse of water in Otay. The task force met for 11 months and developed key foundational components for what would become the Otay water budget program. These components included: Creation of landscape water budgets using historic ET for tall fescue (which amounted to 48" of applied water per year); and a penalty structure for exceeding water budget allocations. Allocation could be reduced to 34" if Otay faced supply constraints, but this situation never occurred while the program was in place.

Official approval for the program quickly followed and Otay's water budget rate structure went into effect on June 1, 1992 – without the penalty component which came into effect three months later in September. Dedicated landscape accounts including golf courses were given 100% of the 48" application rate and customers were required to measure their own properties and provide the information to Otay who followed up with a spot verification program. According to Tubiolo, Otay tried to do as much advance education with customers as possible, but she estimates it took two years to get all of the "kinks ironed out". Re-measuring of landscapes was frequently required during the program.

Otay Water Budget Penalty Structure

The Otay water budget rate structure was designed specifically for landscape irrigation and never applied to mixed-meter customers where indoor and outdoor use were billed together. The rate structure design is shown in Table 4.6 and is significantly different from the water budget structures developed in San Juan Capistrano and IRWD. The Otay water budget was a 12 month allocation, shaped by the historic ET curve for the region. Customers who used less than their allocation in any given month were allowed to bank their un-used allocation for a future month over a 12 month period, a feature not seen in any other water budget rate structure. Customers were allowed to bank up to 25% of their total allocation this way and the banked water would reduce any penalty charges.

The rate structure itself was a modified flat rate or single block rate, where all water use was billed at the same rate. The first time a customer exceeded the monthly allocation, a warning letter from the water utility was sent out with the bill. The second month the allocation was exceeded, the customer was billed a 100% penalty on the overuse. The third month of an exceeded allocation resulted in a 400% penalty on the overuse. The fourth month penalty was 800%.

"This got people's attention," Tubiolo said.

Customers in a penalty situation were invited to contact Otay and request an audit. Audits typically involved verifying the landscape measurement and evaluating the performance of the irrigation system at the site. Tubiolo said she would bring the billing history for the site with her to the site audit and often this would be convincing to the customer. Customers who requested this service would then be granted a three month grace period to bring their usage down within their allocation. If the customer stayed out of the penalty for 90 days then Otay would grant a refund of the penalty dollars collected. According to Tubiolo, only 1-3% of water budget customers ever went into the penalty and only 4 or 5 refunds were ever granted.

Table 4.6 Otay water budget penalty structure (no longer in use)

Landscape Water Use	Penalty Structure
Up to monthly allocation	No penalty. Customer billed for usage.
1st month exceeding budget	No penalty. Customer billed for usage. Warning letter included with water bill.
2nd month exceeding budget	100% penalty on usage above budget. Customer billed at normal rate for allocation and at 2X normal rate for use above allocation.
3rd month exceeding budget	400% penalty on usage above budget. Customer billed at normal rate for allocation and at 4X normal rate for use above allocation.
4th month exceeding budget (and all subsequent months exceeding budget)	800% penalty on usage above budget. Customer billed at normal rate for allocation and at 8X normal rate for use above allocation.
Customers allowed to bank up to 25% of annual budget by using less than the allocation. Banked water is then credited against potential penalties.	

Tubiolo explained that a small percentage of the water budget customers (1.5% or less) would have chronic mismanagement problems that resulted in penalty charges being assessed even after an audit was performed. Schools proved to be one of the most challenging types of sites to manage under this program she said. The City of Chula Vista decided to use a reduced landscape allocation of 32" of water over the year rather than the 48" average selected by Otay. This reduced water budgets in Chula Vista by 33%.

By 1996, a number of homeowner's associations (HOAs) in Otay had fully caught on to the water budget program and began writing water budget compliance into their landscape maintenance contracts. Tubiolo explained that the contracts would specify that any penalty charges due to budget overages would be paid by the landscape contractor. This solved the problem of the accountability disconnect that often occurs in a large landscape account where the person who receives and pays the bill has little or no control over water use at the site.

2000 – Otay Ends Water Budget Surcharges

Not everyone in Otay was as enthusiastic about the water budget rate structure as Jan Tubiolo. According to William Granger, current Water Conservation Manager for Otay, the customer service manager, "still has nightmares" about the program because of the customer complaints it generated. Granger also noted that 70% of the landscape accounts used self-reported landscape area information. Over-statement of landscape area would result in a larger water budget and if widespread could defeat the intent of the rate structure. He also noted that there were complicating factors with adjacent properties and budget determinations.

In 2000 the Otay board of directors voted unanimously to terminate the water budget program and landscape penalty rate structure it implemented in 1992. Board members stated that the program would continue but since the drought was long over the program would shift to have more of an educational focus where customers could be informed if they exceeded their monthly water budget and encouraged to take advantage of Otay conservation programs that highlight potential savings. Jan Tubiolo retired in 2001 shortly after the board's decision.

Between 2000 and spring 2003, the monthly water budget continued to be on the Otay water bills, even though the surcharges were removed. In addition, up until December of 2005, the water budget program was still intact within the Otay Code of Ordinances. Water budget-based surcharges would have been reinstated had Otay entered a water shortage between 2000 and 2005.

In the fall of 2002 through spring 2004 Otay sent separate monthly letters to landscape irrigation customers notifying them if they were over their voluntary water budget and recommending a review of the landscape area. The letter also suggested participation in the Otay large landscape audit program. According to Granger, these letters generated a great deal of interest in the large landscape audit program.

In 2002 Granger reviewed the landscape water budget program and noticed that most of Otay's landscape accounts were well under their water budget. He concluded that either significant deficit irrigation was occurring or many customers had over reported their landscape area.

2005 – Otay Adopts Voluntary Water Budget Program

Otay again moved into the water budget arena in the spring of 2005 with a voluntary landscape water budget program. The voluntary program invited dedicated landscape customers to receive monthly information electronically on the amount of irrigation water that they applied. The new program is informational only and no penalties or charges are assessed to customers.

Conservision, a California-based consulting firm specializing in implementing outdoor efficiency programs, was contracted to run the voluntary program. To date approximately one third of the landscape accounts in Otay have signed up for the program. Of these customers, 25% are from the City of Chula Vista.

Otay continues to recognize the benefits and efficiency potential of a water budget rate structure. They are currently exploring options for re-establishing some version of a water budget rate structure in the future. Otay's revised Code of Ordinances has been recently re-written to allow implementation of a water budget program should the agency enter into a long term water shortage and need to reduce water demand by 10%, 15%, 25% or more.

CASE STUDY: CENTENNIAL WATER AND SANITATION DISTRICT

Located just south of Denver, the Centennial Water and Sanitation District (Centennial) provides water and sewer service to 87,000 people in the Highlands Ranch community in Colorado. With a customer base that is largely residential, Centennial's service area covers 14.1 square miles of developed land. In 2005 Centennial's total treated water demand was 15,230 acre-feet (AF). Build-out demand is estimated to be between 20,000 and 22,000 AF. Centennial operates a dual source water system that utilizes approximately 70% surface water and 30% groundwater in an average year.

In 2003 Centennial became the first water provider in Colorado to implement a water budget-based rate structure. Both the utility and the customer base have been largely pleased with the results. Centennial estimates that demand has been reduced by 25% and revenues have been stable. Customer reaction has been positive and the rate structure has been seen as fair and equitable both across and within classes of customers.

2003 – Water Budgets as Drought Response

During the severe drought conditions in 2002, the Centennial Water and Sanitation District decided to implement a water budget-based rate structure as a method for reducing demand. In 2002 Centennial experienced greatly reduced surface water yields and relied on higher use of groundwater. Centennial desired increased, permanent reductions in water demand to increase system reliability and stretch their water supply. The timing was favorable for introducing a significant new rate approach that was believed to be fair, innovative, and effective in saving water. The rate structure was officially adopted May 1, 2003.

"The drought gave us political cover," said Rick McLoud, Water Resources Manager for Centennial, "This could be seen as a kind of assertive thing for government to do."

McLoud learned about the water budget concept at a talk by Tom Ash, formerly of Irvine Ranch Water District. Ash was subsequently invited to come and meet directly with Centennial leaders and staff. As a result of these meetings and research by Centennial's financial managers, political leaders and Centennial staff were supportive from the outset and quickly became convinced that the water budget approach was the "right thing to do".

Formulation of Centennial Water Budget Rate Structure

The Centennial water budget-based rate structure was developed to cover all customers and customer classes in the service area – single-family and multi-family residential, commercial, and dedicated irrigation accounts. The rate structure has two major components: (1) a fixed water service availability fee; and (2) a variable water consumption rate component.

Using the IRWD model, Centennial chose to set an increased fixed fee of $12.50 per customer per month (previously the fixed fee was $4.95 per month). This was an intentional choice made to assure revenue stability for the District.

The variable water consumption rate component consists of a 4-tier rate structure where the break point between tiers is based on a water budget, developed for each individual customer in the district. The Centennial rate structure tiers are shown in Table 4.7.

Table 4.7 2006 Centennial Water and Sanitation District Rates

% of Budget	Rate ($/kgal)
0 – 100%	$2.50
101 – 120%	$3.45
121 – 140%	$5.20
140%+	$7.80
Fixed monthly fee	$12.50

The tiers and rates were somewhat different when first implemented in 2003 as part of drought response. The top tier was set at $16.00 per kgal. Adjustments were made to the width of the tiers and the rates based on feedback from customers and revenue requirements.

"One of the things we like about this rate structure is that the various components can be adjusted," explained McLoud. "We have tweaked the system in various ways since it was first implemented. It's flexible."

Residential Budgets

Residential water budgets are formulated by constructing an indoor and an outdoor component. The indoor component is 6,000 gallons per month per customer based on an assumed 65 gallons per capita per day for a family of three. Customers with large families can sign an affidavit and receive an additional indoor allotment. Budgets are calculated on a bi-monthly basis since residential customers are billed bi-monthly.

Outdoor budgets are based on the actual lot size (obtained from tax assessor records) multiplied by an irrigated area factor of 45% and 27 inches of water for the year applied across that area. In the Centennial system all customers are provided 45% of their total lot size as the irrigated area. For example, a 10,000 square foot lot would receive an annual outdoor water budget of 75.7 kgal (10,000 sf x 0.45 x 27 inches ÷ 12 inches/ft x 7.48 gal./cubic foot). Outdoor budgets are distributed through the irrigation season (April – October) based on the historic evapotranspiration curve in the Highlands Ranch area.

A key component of the Centennial outdoor budget, the 45% irrigated area factor, was developed through empirical research in the service area. A sample of approximately 1,000 accounts was taken and a detailed irrigated area determination was made using aerial photography and GIS technology. An example of this analysis is shown in Figure 4.3.

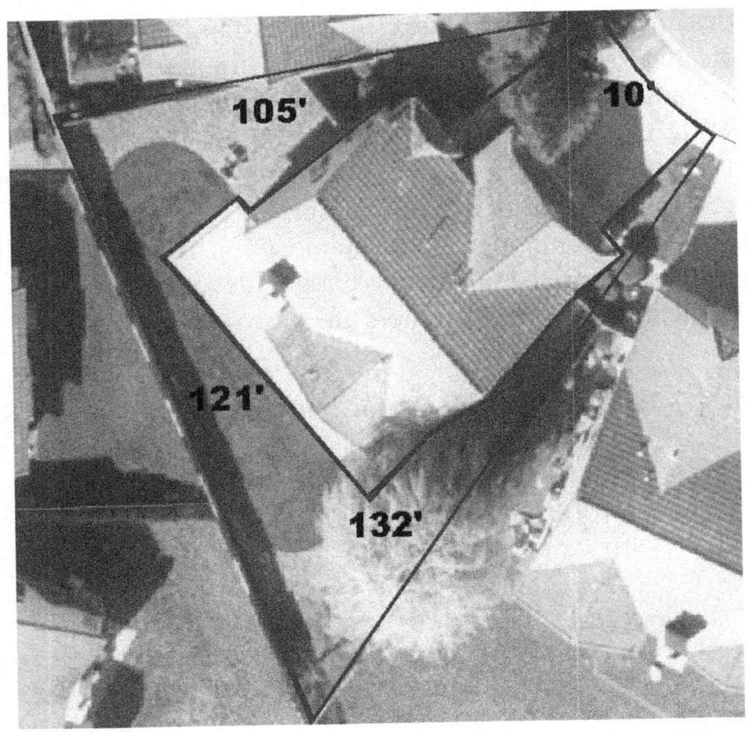

Figure 4.3 Aerial photo analysis utilized by Centennial

Centennial's analysis of irrigated area showed that on average, properties in the service area consisted of 45% irrigable area and 55% impervious area (house, driveway, and hardscape). Rather that perform this analysis for every single customer in the system, Centennial adopted the 45% irrigable factor into the rate structure and applied it to all residential customers. The utility board also decided that there would be no formal appeals process for landscape water budgets so customers would simply have to accept the allocation provided. Figure 4.4 shows an example of a Centennial bi-monthly residential budget over the entire year.

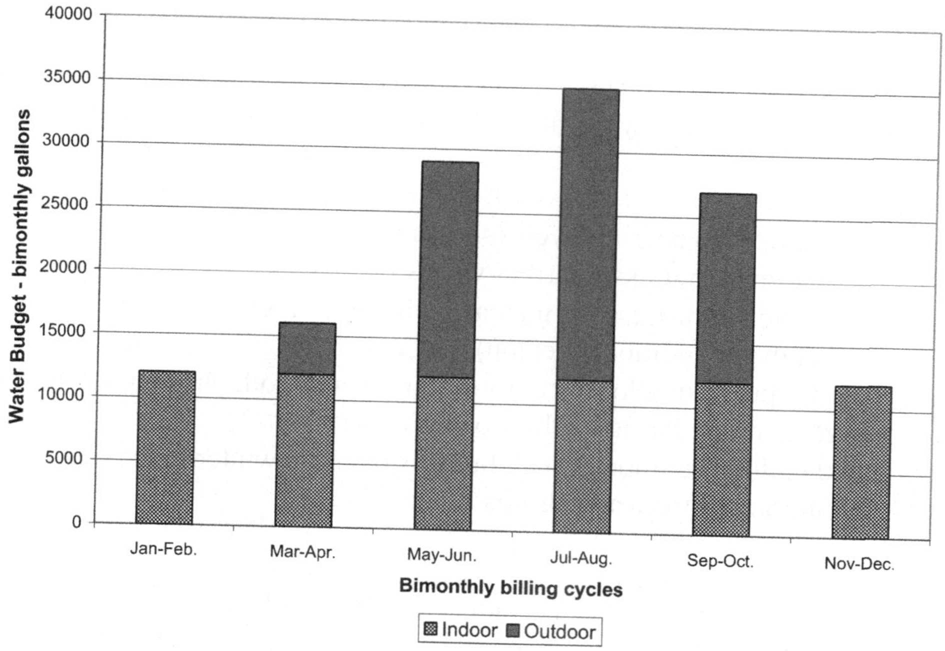

Figure 4.4 Typical annual residential water budget in Centennial Water and Sanitation District

Commercial Budgets

Water budgets for commercial customers in Centennial are similar to those developed for the residential sector and are also based on an indoor and an outdoor component. Indoor budgets are based on the historic use for each commercial customer. Additional indoor factors are considered on a case by case basis.

Outdoor use is based on the actual irrigated area at each commercial site. The customer is responsible for submitting this data to the utility. A lower outdoor budget is provided until the customer responds with a submission. As with residential customers, 27 inches per year is provided for outdoor irrigation.

Although the commercial sector in Centennial is relatively small, developing accurate and reasonable indoor budgets has been a challenge. Indoor use in the commercial sector is variable by season and business. In the three years since the water budget program was implemented the commercial sector's total demand has been consistently above the total budgeted demand. McLoud stressed that Centennial is working to develop a more fair indoor budgeting methodology that will nudge commercial customers to be more efficient but will not unfairly penalize them for legitimate use.

Adjustments to Water Budget Program

In September 2003, just a few months after Centennial's water budget program was initiated several key adjustments were made to improve performance of the system. These included:

- Reducing block rates to current levels (shown in Table 4.7)
- Adjusting block break points to current values (shown in Table 4.7)
- Including the affidavit process for increasing indoor budgets for larger families (1,500 gallons per month per additional person)
- Adding a permit program allotting extra water for new sod, but only in the spring and fall (to discourage planting during the heat of summer)
- Adding 1,000 gallons per month to all budgets over the winter months to allow for wintertime watering of trees and shrubs.

With these changes Centennial's rate structure has been in place for over three years and no additional changes have been made. The adjustments described above illustrate the flexibility of the Centennial water budget system, a feature that McLoud was particularly pleased about.

Implementation Process

Implementation of the Centennial water budget-based rate structure was a quick and fairly inexpensive process. The entire effort was accomplished in six months using primarily in-house staff. Obtaining lot size information from available electronic county records was a key to success McLoud said. Seasonal evapotranspiration rates were obtained from local weather stations in the service area and this helped to establish 27 inches as a reasonable amount of water to provide customers for outdoor irrigation.

The computer billing system in Centennial did not need to be replaced to implement the new rate structure. Modifications were made to the existing software to accommodate the lot size data and budget calculations. This process was accomplished quickly so that the program could be implemented in a timely fashion.

Communication with customers about the new rate structure was another critical element of the implementation process. Centennial conducted workshops and public meetings and participated in community events to publicize the new rate structure. An informational postcard explaining the rate structure was sent to every customer in the service area and detailed information was made available on the utility web site. The local newspaper published articles about the change and customer service representatives were trained to handle calls from customers looking for answers about the system.

Centennial has an on-going communication and education program associated with the rate structure. Prior to each irrigation season a bill insert is sent to every customer reminding them of the water budget system. Centennial has also developed packages of information for new customers to explain the rate structure and answer the typical questions that arise from unfamiliarity with a water budget system.

A sample water bill from Centennial (Metro Districts of Highlands Ranch) is shown in Figure 4.5. The bill shows the current month's water budget and the budget for the upcoming month. The bill shows fairly limited information about the rate structure, but provides information on where to obtain details and direct questions.

METRO DISTRICTS
Highlands Ranch
62 West Plaza Drive • Highlands Ranch, CO • 80129

Please make Checks Payable to HRMD

Account#: [redacted]

Due By: 12/23/06

Total Due: $65.89

[redacted]
HIGHLANDS RANCH CO 80129-4397

[redacted]

PLEASE RETURN THIS STUB WITH YOUR PAYMENT. RETAIN BOTTOM PORTION FOR YOUR RECORDS.

Metro Districts of Highlands Ranch
62 W. Plaza Drive, Highlands Ranch, CO 80129

Visit our Web site at ***www.highlandsranch.org*** for information regarding the Metro Districts, board meetings, water conservation, rates and updates. Office hours are Monday through Friday, 8 a.m. to 5 p.m.

IMPORTANT TELEPHONE NUMBERS

Billing Inquiries:	303-791-0361
General Information:	303-791-0430
Water Emergencies: (ext. 545)	303-791-2185
Street Light Repairs: (Xcel Energy)	1-800-895-1999
TTY:	1-800-659-2656

Customer: [redacted]
Service Address: [redacted] HIGHLANDS RANCH, CO

Description	Read Dates From	Read Dates To	Readings in 1000 gallons From	Readings in 1000 gallons To	Usage In Gallons	Prior YR Usage Same Period	Cur Period Budget	Next Period Budget	Amount Due
Street Light									4.30
Wastewater									25.09
Water Availability									25.00
Water Usage	10/04/06	11/22/06	771	776	5000	6000	16000	13000	11.50
Usage up to 100% of current budget			5,000 x	2.30/1,000 =	11.50				

Total Due By 12/23/06	$65.89

When the total balance is not received by the due date, the account will be considered past due and a penalty of $5.00 will be imposed.

The factors used to determine your water budget are one or more of the following: 1) average Highlands Ranch indoor consumption; 2) your lot size as obtained from Douglas County Assessors Office; 3) targeted landscapable area per lot based upon average house footprint and other factors; 4) sufficient irrigation water to maintain landscaping.

Other Messages: Your next read date will be the week of January 22nd-26th, 2007. Remember to winterize your sprinkler system to avoid breaks from freezing and possible leaks next spring. No outdoor irrigation is allowed between 10:00 am and 6:00 pm year-round.

Serving You Every Day

Figure 4.5 Sample water bill from Metro Districts of Highlands Ranch, Colorado

Impact of Centennial Water Budget Program

Centennial implemented its water budget-based rate structure in the hopes of reducing water demand and stretching their water supply. Overall the Centennial staff has been pleased with the results and has documented an average 25% reduction in demand after implementing the program. Figure 4.6 shows actual demand from 1993 – 2006 and compares this against forecast demand based on actual population increases and demand at pre-2002 levels. Even though Centennial continues to add customers at a fairly steady rate, its demand has decreased 18 – 31% starting in 2003 when the water budget-based rate structure was implemented. These savings are similar to the reductions reported by Irvine Ranch and San Juan Capistrano, and if sustained, represent a substantial benefit to Centennial as it continues to grow to build-out.

The Centennial staff believes that much of the water savings achieved through the rate structure change involves reduction in water waste. Customers who had previously not paid much attention to their irrigation practices have received a powerful message and it has had a measurable impact. Analysis of consumption trends in Centennial showed McLoud and his staff that most of the reduction in water use is coming from customers who previously had the highest demand. Customers who exceed their budget substantially paid $16 per kgal in the highest tier in 2003 before the rates were reduced to post-drought levels. Complimentary conservation programs and measures banning watering during the heat of the day and prohibiting wasteful irrigation practices have worked in tandem with the rate structure.

Other outcomes for Centennial include improved stability of the revenue steam. This is due in large part to the intentional decision to increase the fixed fee portion of the bill from $4.95 to $12.50 with the water budget rate structure.

"We hedged our bets by putting a higher fixed charge," said McLoud.

The variable component of the bill is still significant to most customers, particularly those who exceed their water budget by 20% or more. Centennial's revenue projections do assume that some customers will exceed their budgeted amount and land in the higher tiers, but this type of assumption is typical of agencies that implement an increasing block rate structure. The overall impact of the water budget-based rate structure has been to provide a stable and sufficient revenue stream, even as demand has reduced because of the program.

"The water budget system is equitable and understandable to customers," McLoud explained. "It rewards efficiency and strongly discourages waste. It also preserves choice in that customers can use more water if they want to. They just pay a higher rate.

"This system is effective at conserving water with or without a drought and provides substantial incentive for customers to change old and inefficient landscaping approaches."

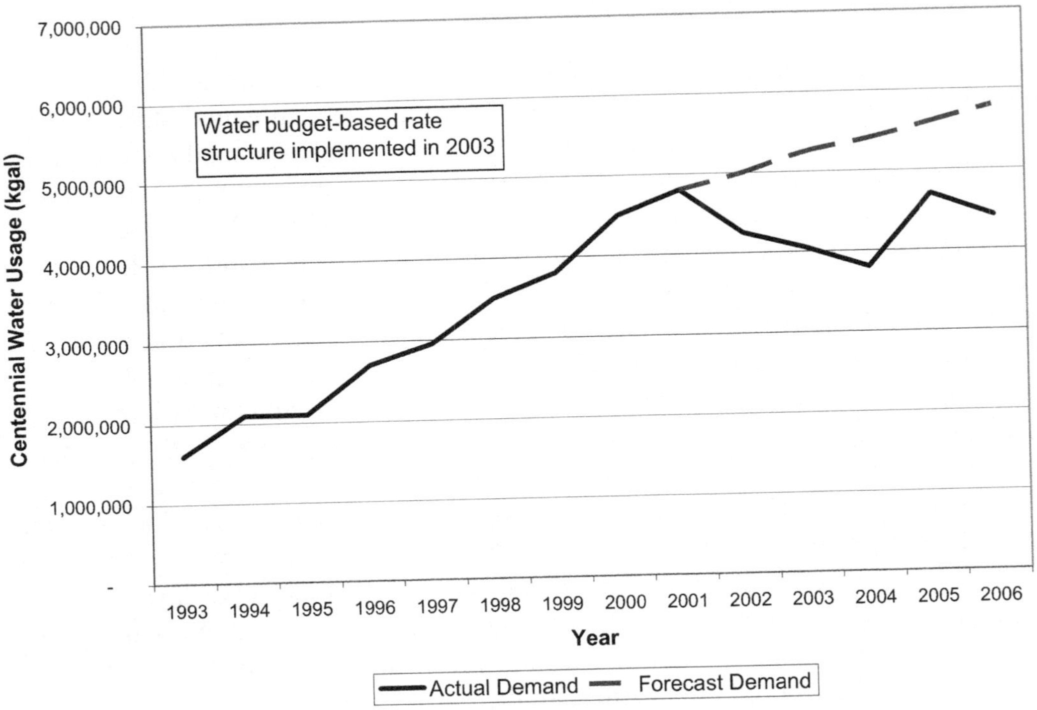

Figure 4.6 Actual and forecast demand in Centennial Water and Sanitation District

CASE STUDY: CITY OF LOS ANGELES DEPARTMENT OF WATER AND POWER

The Department of Water and Power (DWP) serves the City of Los Angeles and some small adjacent areas. The City's increasing block rate evolved out of a Citizens Blue Ribbon Committee in 1993 and was widely noted for several innovative features—a seasonal increasing block rate structure based on marginal cost pricing principles, elimination of fixed charges (resulting in a 100 percent commodity-based revenue generation), an intensive home survey program that accompanied the roll out of rates, a quarterly rate adjustment mechanism to balance revenue, and an ongoing public involvement and outreach program (Hall and Hanemann, 1996), (Chesnutt et. al. 1996). It was this latter public involvement process that directly resulted in the evolution from a generic block rate structure—where each block is defined by a fixed amount of water—to adoption of a rate structure in 1995 that uses information on lot size, weather zone, and household size to define block size.

The DWP rate structure is a success story in that it helped flatten water demand in spite of a growing population and economy. Figure 4.7 below, taken from DWP's 2005 Urban Water Management Plan illustrates the combined effect of their water budget-based rate structure and a portfolio of innovative water use efficiency (conservation) programs. DWP's rate structure has maintained required utility revenues, reinforced incentives to use water efficiently, and has achieved customer acceptance within a major metropolitan area having a heterogeneous population.

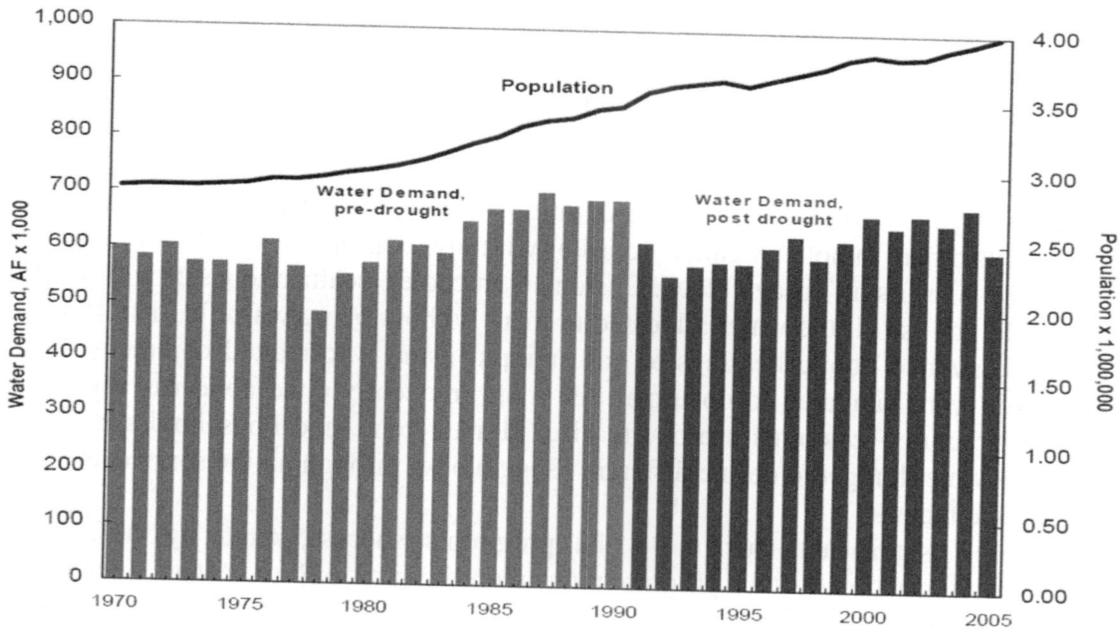

Figure 4.7 Water demand and population growth in Los Angeles[11]

Overview

The DWP service area receives approximately 12 inches of precipitation in a normal year. The DWP obtains 30 percent of its supply from surface sources and 15 percent from ground sources (with up to 30 percent during droughts), with the residual being purchased from the Metropolitan Water District of Southern California. The DWP has a service area of 469 square miles and serves a population of almost 4,000,000. It has approximately 480,000 residential customers, 120,000 multi-family residential customers, 60,000 commercial customers, 6,600 industrial customers, and 7,000 governmental customers. About two-thirds of total water consumption is accounted for by residential customers, one quarter by commercial and governmental sectors, and four percent was served to industry. (The remaining amount is used for fire-fighting, or lost through evaporation, system leaks, and breaks).

Implementation Process

DWP's rate structure also developed in response the California drought of the late 1980s and early 1990s, similar to IRWD, San Juan Capistrano, and Otay. The voluntary and mandatory curtailments during the 1990 to 1992 period, increased concern about water reliability, and the need to promote water conservation through rates were cited as reasons for rate reform. A Blue Ribbon committee recommended adoption of a seasonal increasing block rates structure in 1993 after more than a year of work. The impetus for rate revision came both from the lingering drought and a need to address the issue of how to fairly allocate shortage. The environmental community was quick to point out that customers who conserve should not have to be burdened

---

[11] http://www.scag.ca.gov/rcp/pdf/uwmp/LosAngeles/LADWP_2005UWMP.pdf

with a higher price or imposed curtailment, correctly observing that percentage curtailments unfairly punish those customers who have already invested in conservation measures.

Interestingly, water budgets were considered at this time but were set aside due to feasibility concerns. A series of public workshops were held after the adoption of the increasing block rate. Feedback from customers participating in these workshops resulted in reconsideration of the application of a single usage block definition to DWP's diverse residential customers. Specifically, customers in hotter weather zones or with bigger properties said they could not stay within the "one-size-fits-all" usage block. Customers in cooler weather zones or with smaller lots said they could not use all the water allocated to them without being wasteful. The Blue Ribbon committee reconvened public hearings to gather input using a formatted comment card. The comment cards were tabulated to document the most frequently cited concerns about the then existing rate structure—customer's property size, climate zone, and household size. Thus the idea of using a water budget to define different usage blocks throughout the service area returned using these three factors.

The most significant barriers to the program's success, cited by staff, were the complexity of information system programming, reconfiguring billing system reports, and the address matching needed to obtain accurate lot size information. These barriers were overcome in a six to eight month time frame through the concentrated efforts of existing employees. In general, customers responded very well to the revised 1995 rate structure that incorporated five categories of lot size, household size, and temperature zone (3 zones). Of these three factors, staff cited the lot size adjustment as the most successful in that customers recognize and understand it, and contributed most to acceptance of two tier pricing with a large rate differential between tiers. According to DWP staff, the household size adjustment and temperature zone adjustments have been less noticed by customers.

The 1995 rate structure implemented in DWP (and still in place) is a two tier seasonal increasing block rate structure that applies to all customer classes. The first tier rate includes a revenue adjustment factor that assures that a targeted minimum revenue will be collected. It also includes pass-through adjustment factors for water procurement, water quality improvements, security, Owens Valley regulatory and low-income subsidies. The second tier rate is based on marginal cost and has a seasonal component. Peak second tier rates are applicable to usage in June through October; off-peak second tier rates are applicable to usage in November through May. The formula for determining first tier usage for peak and off-peak periods varies with customer class.

For single-family residential customers, the breakpoint between first and second tier usage is based on lot size (five categories), temperature zone (three zones), and household size (the household adjustment involves a sliding scale of allowed extra first tier usage for households of seven persons to households of thirteen or more). For multi-family residential customers, all usage in the off-peak period is billed at the first tier rate; usage in the peak period in excess of 125 percent of the previous average winter usage for December through March is billed at the second tier rate. A similar rate structure is applicable to commercial and industrial customers, as well as to governmental customers. However, non single-family residential customers with high seasonal variation in usage as well as large turf customers that have achieved maximum practical reductions in usage via conservation practices may apply to have 95 percent of their usage in the peak period billed at the first tier rate. The Department of Public Works of the City of Los Angeles sets wastewater rates; the sewage charge is generally based on metered water usage, with a standard exemption for water not returned to the sewers (i.e., outdoor usage). For residential customers with up to four dwellings on a single meter, the sewerage charge is based on each customer's lowest winter water use. A private water submetering program is also offered, whereby a customer many install one private submeter per water meter to separately measure outdoor water use. The outdoor usage is then exempted from the sewerage charge.

DWP also has a lifeline rate available to low-income or senior/disabled customers. Income eligibility is determined by an adjusted gross income (AGI) threshold, currently set to a household AGI of twenty thousand five hundred dollars per year or less. To qualify, customers submit a lifeline rate application to the city clerk with a copy of the previous year's tax return for verification.

Billing System and Software

Implementation of the DWP water budget rate structure did not require new billing software. Instead, programming to the existing system was accomplished and new billing forms were created using an estimated 4 to 5 FTE for a period of six months. DWP used publicly available tax assessor data to define lot size. Zip codes were used to define temperature zones. Since the initial implementation, the DWP water budget billing system has not required any additional or new staff to operate. Total costs to develop both the 1993 and 1995 rate structures were estimated to be about $750,000 ($300,000 in direct consulting and expert costs and the remainder in customer in-kind contributions of time.) Staff pointed out that this amount worked out to be about $0.68 per customer and felt that the time and expense were well worth the rate reform enacted.

A sample water bill from DWP is provided in Figure 4.8.

12 Years of Operational Experience

The DWP rate structure has been in place for more than a decade and continues to be an effective billing methodology and water conservation tool for the Department. Staff ranked its effectiveness as a "9" on a ten point scale.

LA DWP

Los Angeles Department of Water and Power
P.O. Box 30808, Los Angeles CA 90030-0808 Web site at http://www.ladwp.com

[redacted]

PAGE 1

[redacted]

[redacted]

Bill Issued 01/12/07 Due Date 02/02/07
Amount Of Previous Bill $ 221.69
Payments Since 11/08/06 $ 221.69-

DWP ENERGY SERVICES- 1(800)342-5397
This Bill Covers 11/06/06 To 01/10/07.
Energy Used 892 KWH* Meter 02-Electric Total $ 94.17

$ 1.45 is your daily average cost for energy.

DWP WATER SERVICES- 1(800)342-5397
This Bill Covers 11/06/06 To 01/10/07.
First Tier 22 HCF**
Total Water 22 HCF Meter 01-Water Total $ 47.62

$ 0.73 is your daily average cost for water.
Your cost per gallon is less than 1/2 cent.

DWP SUBTOTAL $ 141.79

METERING INFORMATION	02-ELECTRIC	01-WATER
Meter Number	9-967787	50407640
Current Read	18130	3705
Previous Read	17238	3683
Constant	1	1
This Year-Use	892 KWH	22 HCF
Days Billed	65	65
Daily Average	14 KWH	253 GAL
Last Year-Use	784 KWH	22 HCF
Days Billed	66	66
Daily Average	12 KWH	249 GAL

*KWH(KILOWATT HOUR) IS 1,000 WATT HOURS ** 1 HCF = 748 GALLONS

Figure 4.8 DWP sample water bill

Recommendations for Other Agencies

DWP staff provided a few recommendations to agencies considering implementing a water budget rate structure. These include:

- Public acceptance is crucial. The public workshops allowed DWP to elicit customer feedback on the 1993 increasing block rate structure, construct adjustments to the rate structure, and win over more of its customer base.

- It is critical that the water budget have a rational basis that can be clearly communicated to customers. By providing a rational basis for defining the width of usage blocks, water budget-based rate structures are perceived by customers as being intrinsically fair.
- The water budget must address customers' primary concerns.

CASE STUDY: CITY OF BOULDER

The City of Boulder, Colorado is located along the Front Range of the Rocky Mountains, approximately 35 miles northwest of Denver. The estimated service population of the water utility was over 108,000 persons as of 2000 and projected to grow to approximately 126,000 by build-out (Integra, 2000). Boulder also supplies water to a significant number of commercial, industrial and institutional customers including the University of Colorado. As of 1995 there were a total of 85,400 jobs in Boulder's service area with projections to 115,200 jobs at build-out.

The City obtains water to supply its customers from two major water sources: (1) the watersheds of Middle and North Boulder Creeks, which lie on the east side of the continental divide, and (2) water taken from the upper reaches of the Colorado River through the Colorado Big Thompson and Windy Gap projects. Both of these supply sources rely on snowmelt for diversion and storage. Water from these sources, in combination with the City's reservoirs and exchange rights, provide a water supply that was estimated to have a firm yield (based on the 1954 drought) in excess of the estimated dry year demand of 26,800 acre-feet. The estimated firm yield of the raw water supply was estimated as somewhere between 33,000 and 37,000 acre-feet (Hydrosphere, 1999). Until 2002, 1954 was the driest on record in the region, and was commonly used by water planners as the design drought for system reliability studies.

Drought and Water Budgets

A water budget rate structure was first contemplated in Boulder in 1999 as part of a water conservation futures study (Hydrosphere, 1999). There was little movement on the idea until 2002 when the entire region experienced one of the driest years in more than 100 years of measurement. By May 2002 the snow pack in Boulder's watershed had runoff or evaporated. The water supply emergency led to a series of mandatory water use restrictions during the summer of 2002. The restrictions were effective in reducing water use by more than 20% overall, but landscapes suffered and some citizens were unhappy with the drought enforcement policies. This led to a useful discussion by the customers, utility staff, and City Council on how future droughts might be managed more effectively. The idea of a water budget rate structure was re-introduced at this time.

The idea of a water budget rate structure was not a drastic departure from the way that the City had billed for water for the past 20 years with a three tier increasing block rate structure where the first tier was sized by the average winter consumption (representing primarily indoor use) of the customer. Coincidentally, Boulder's water billing computer system and software were outdated and needed to be replaced at this time. This provided an opportunity for Boulder to reconfigure the water rate structure at the same time.

With the support of the Water Resources Advisory Board, an advisory board to the City Council, in December 2002 the Boulder City Council approved funds for a feasibility study to examine options for adopting a new rate structure that could include water budgets. The water budget concept received strong support from citizens and the Water Resources Advisory Board. In December 2004 , Boulder City Council adopted a five tier increasing block water budget rate structure. The actual water budget billing did not start until January 2007 after implementation of the new computer billing system.

Calculation of Rates

Boulder's water budget structure is a five tiered increasing block rate in which the breakpoints between the blocks are based on each customer's monthly water budget. The budget is intended to represent the amount of water that the customer requires for normal domestic purposes; to care for a landscape, operate a business, etc. in a reasonable manner under locally accepted standards and expectations. For example, landscaping patterns in Boulder have traditionally not been dominated by broad expanses of turf, so the landscape budgets were designed to accommodate landscapes with mixtures of turf and non-turf plant materials, rather than full turf landscapes. The five block water budget rate structure that is used by the City of Boulder is shown in the Table 4.8.

Table 4.8 Water Rates for City of Boulder 2007

Tier	**Rate** (per kgal)	**Use** (% of water budget)	**2007 Charges** (per kgal)
1	¾ base rate	0 – 60%	$1.88
2	base rate	61 – 100%	$2.50
3	2 x base rate	101 – 150%	$5.00
4	3 x base rate	151 – 200%	$7.50
5	5 x base rate	201% +	$12.50

Establishment of Water Budgets

Single-Family Residential

Boulder establishes monthly water budgets starting from an annual allocation for each customer that is shaped to a monthly curve. Boulder's water budgets are individualized, but a different budget development methodology is employed for each of the major customer categories. For single-family residential customers the indoor budget is fixed at 7,000 gallons/month per customer which provides 60 gcd for a family of 4. Additional water is provided to accommodate some winter irrigation of trees and shrubs that is frequently required during dry/windy conditions in Boulder. Allocation adjustments are available upon request for customers with special needs or requirements. Outdoor budgets for single-family residences are based on the landscaped area, and provide 15 gallons/square foot (gpsf) for the first 5,000 sf of landscaped area; 12 gpsf for the next 9,000 sf; and 10 gpsf for everything above 14,000 sf. Ten gpsf in Boulder is the water requirement for low water use native plants, so this approach is designed to discourage wall to wall turf on large landscapes. Irrigated areas were measured using Boulder's geographic information system (GIS). Annual allocations are allotted into monthly budgets based on historic climate data.

Table 4.9 shows the annual outdoor allocation (budget) for a single-family home with a measured landscape area of 14,400 square feet. The annual outdoor allocation is distributed throughout the year to meet changing monthly seasonal outdoor watering needs. Table 4.10 shows the percentage by month that is applied to the annual outdoor allocation. These percentages were derived from historic climate data.

Table 4.9 Outdoor allocation example, Boulder, CO*

Irrigable Area (square feet)	Gallons per Square Foot (gpsf)	Annual Allocation (gal)
5,000	15	75,000
9,000	12	108,000
400	10	4,000
Total Annual Outdoor Allocation		187,000

*Example assumes a single-family home with a 14,400 sf landscaped area

Multi-Family Residential and Dedicated Irrigation Accounts

Multi-family allocations in Boulder provide 4,000 gal/month for each dwelling unit for indoor purposes. Irrigation allocations for multi-family accounts are established using 15 gpsf per year for the entire irrigated area. Allocations for irrigation-only accounts (dedicated irrigation customers) are also established using 15 gpsf over the entire irrigated area. The rational for providing multi-family and irrigation accounts 15 gpsf for all irrigated areas, and not reducing the application rate as is done for single family accounts, is that these landscapes are typically used by the public for recreation. This potentially more intensive use combined with the likelihood of more turf grass justified the increased allocation.

Table 4.10 Monthly outdoor allocation percentages, Boulder, CO

Month	Percent of Annual Outdoor Allocation
January	0%
February	0%
March	1%
April	7%
May	14%
June	20%
July	20%
August	18%
September	12%
October	7%
November	1%
December	0%
Total	**100%**

Commercial, Industrial, and Institutional Customers

Boulder recognized that setting budgets for commercial, industrial and institutional (CII) users is much more complicated than for other customer categories due to the variable nature of CII water use. In order to allow the water budget system to proceed without delay while data were collected for the CII accounts, the City elected to use the historical use of CII customers to set their initial budgets. Each CII customer had its initial water budget set at 100% of the water use during 2005. Monthly budgets were created as the average of the total annual use in 2005 (e.g. total use in 2005 ÷ 12). This approach may pose problems for CII customers with a strong seasonal component to their demand.

Implementation of Boulder's Water Budget Rate Structure

The Boulder City Council approved implementation of a water budget rate structure in December 2004, but the existing computer billing system could not accommodate the new rate structure. While the City purchased, tested and implemented (May 2006) the new billing system, the City operated using the old rate structure until January 2007. There were several parallel activities that occurred in 2006 involved with implementing the new structure:

- Focus groups and outreach surrounding new bill design.
- Several additional staff persons were hired to assist with the implementation and transition to the new billing system and rate structure.
- An intensive effort was made to obtain total lot size and irrigated areas for each single-family, multi-family, and dedicated irrigation customer using the City's geographical information system (GIS). Data for each account was stored electronically with other customer information maintained by the billing department.
- Date used in establishing budgets (irrigable area, account type) was forwarded to Red Oak Consulting, who developed individual budgets for all of the city's customers and applied historical water usage information to evaluate rates and revenues. Public meetings were held on rate options.
- A variety of informational fliers were included in monthly water bills for each month between May 2006 through February 2007. One example was titled "What is a Water Budget?" and it explained what the budgets are, how they are determined, and how water use will be billed for consumption in each of the blocks. It provided both a telephone number and web address for customers wishing to obtain additional information. All bill inserts were presented in both English and Spanish. Another example was entitled, "Water Budgets are Coming Soon". This flier explained when the draft budgets would be sent out, how to appeal errors, and offered the customer a water monitor device they can use to track their water use real time. Ideas for water conservation were included with the flier.
- The City arranged with the company providing its automated meter reading (AMR) devices to obtain remote reading devices that each customer could use to obtain real time meter read data. This small device is meant to be mounted on the refrigerator with magnets. The devices are coded to the individual meter and they have three options: the current register reading on the meter and two user defined interval readings that provide the volume used since the last time the register was reset. (This

is similar to the trip registers of most automobile odometers.) These devices are sold for $200, which is billed to the customer's water account. This cost includes both the cost of the monitor device and the cost of updating the meter transponder so it is compatible. As of January 2007, only 20 of the devices had been sold,[12] but increasing sales are anticipated as more customers become aware that they are available.

- A dedicated phone number was provided for customers to call with questions about the water budget billing system.
- Extensive use was made of the City's web site to provide information on the entire water budget system from inception through implementation.[13] The web site describes the billing system, gives answers to frequently asked questions, explains how the bills will look and how to read them, and provides links to several related links on the history, rules and implementation of the program.
- During November, 2006, customers in Boulder were mailed a notice showing proposed monthly water budgets for 2007 compared to their historical use in the prior 12 months. The notices explained how the budgets were derived, and how they can be appealed for adjustment.
- The Public Works department conducted a variety of other outreach methods in an effort to get the word out to customers prior to the debut of water budgets. This included press releases and media relations; materials handed out at farmers' markets, recreation centers and other public events; public meetings to discuss related rate options; memos to WRAB and Council; focus groups and outreach for the development of the new bill design; new Web site content development, including an interactive water use calculator; listserv for water conservation tips and news updates; newspaper ads, and more.
- Forms were developed for requesting adjustments of the proposed water budgets. These forms were provided on-line, and allowed customers to state the reasons for the adjustment and give the corrected parameters for calculating the budget. The forms note that they are subject to audit. As of January 2007 approximately 200 requests for adjustment, out of 30,000 budget notices sent, had been received.[14]
- In January, 2007 the City published a set of rules for implementation of the water budget rate structure for public review and comment.
- The total amount spent in implementing the City of Boulder's water budget program has not been tallied, but if the billing system is included (it would have been purchased regardless of the rate structure implemented) along with various studies commissioned by the City during the process, the cost exceeds $1,500,000.

A sample water bill from Boulder is provided in Figure 4.9. Since Boulder's water budget rate structure had only been implemented for a few weeks at the time this report was prepared, the impacts on water use in the City were unknown. This is an important question for further study.

---

[12] Information provided by City of Boulder Utility Billing office.
[13] www.boulderwater.net
[14] Information provided by City of Boulder Utility Billing office.

Customer No.: [redacted] **Account No.:** [redacted] **Service Location:** [redacted]

Customer Class: Single-Family **Inside/Outside:** Inside **Meter Size:** 3/4"

Reading Date: 09/07/07 **Present Meter Reading:** 48 **Previous Meter Reading:** 35 **Days Billed:** 30

Water	Rate/ 1,000 gal	Used 1,000 gal	Cost
Service Charge			8.55
0 - 11,000 gal	1.88	11	20.68
11,001 - 17,000 gal	2.50	2	5.00
17,001 - 26,000 gal	5.00		
26,001 - 34,000 gal	7.50		
34,001 +	12.50		
Total Water		13	$34.23
Wastewater Service Charge			0.74
Wastewater Quantity		4	14.00
Stormwater/Flood Management			6.75
Total Current Charges			$55.72
Please Pay This Amount			$55.72

Water Budget VS. Actual Use

Water Budget does not carry over from billing cycle to billing cycle.

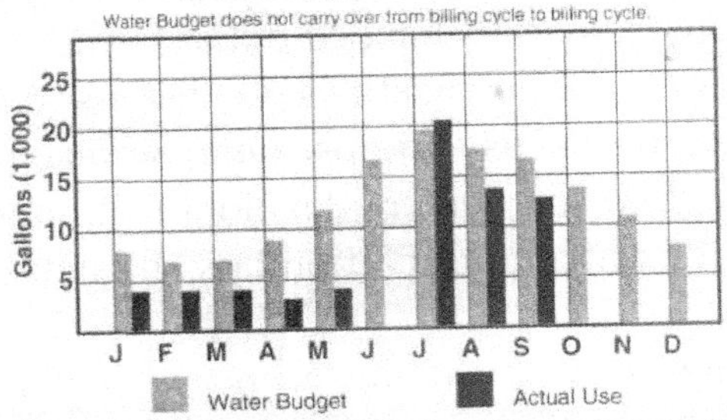

Account Summary

(1,000 Gallons)

Current use: 13 **Budget this bill cycle:** 17

Use last year:

Estimated water budget next bill cycle: 14

Avg. monthly winter usage (Dec.-Mar.): 4

Please Note:

PLEASE DO NOT PAY
TOTAL AMOUNT DUE WILL BE TRANSFERRED AUTOMATICALLY ON OR AFTER 09/25/07

PROPOSED 2008 UTILITY RATE INCREASES

WATER 4%

WASTEWATER 3%

STORMWATER 3%

These proposed rate increases will be discussed by City Council in October as part of the 2008 Budget. For more information go online to boulderutilites.net.

** THANK YOU FOR YOUR PROMPT PAYMENT **

Billing Questions: 303-441-3260
Emergencies: [redacted] 303-413-7100
After Hours Emergencies: 303-441-3249
For more details about your bill: www.boulderwater.net

For automatic payment, contact: 303-441-3260

Esta es informacion importante. Si no la pueden leer, necesitan que alguien se la traduzca.

CITY OF BOULDER

PLEASE DETACH THIS PORTION AND RETURN WITH YOUR PAYMENT

Customer No.	Account No.	Due Date	Balance Forward	Amount Due	Amount Enclosed
[redacted]				DO NOT PAY	$

Make checks payable to: City of Boulder
Please write account number on check

Service Location: Meter Size: Reading Date: Days Billed:

170-10

[redacted]
BOULDER CO 80304-2212

CITY OF BOULDER
Utility Billing Office
Dept 0275
Denver, CO 80263-0275

Figure 4.9 Sample water bill from Boulder, CO

CASE STUDY: SANTA BARBARA, CALIFORNIA

Santa Barbara, California is a city of over 90,000 people with minimal potential for additional growth. It serves almost 25,000 connections through its municipally owned water system. Santa Barbara's current water budget based rate system, which only impacts dedicated irrigation accounts, was adopted in 1989, with full support of its elected leaders, planning and water commissions. Other customers are billed for their water use with a typical three tiered block rate structure.

The goals of the water budget rate structure were fairness, improved long term water management, and short term management of water demand during droughts. Moving from a uniform rate structure, where all customers paid $1.02 per hundred cubic feet (hcf) for all metered water, to an allocation system was fairly smooth and did not involve an extensive public involvement program. Utility bill inserts and newspaper articles provided the tools for informing the public. The current rate schedule for Santa Barbara is shown in Table 4.11.

Table 4.11 Santa Barbara, CA water rates

Customer Class	Block Definition and Commodity Charge
Single Family Residential	1st 4 hcf @ $2.47/hcf Next 16 hcf @ $4.14/hcf All Other @ $4.37/hcf
Multi-family Residential	1st 4 hcf per dwelling unit @ $2.47 Next 8 hcf/dwelling unit @ $4.14 All other @ $4.37
Commercial	100 percent of base allotment @ $4.14 (base allotment = avg. monthly consumption during most recent January to June period). All other @ $4.37
Industrial and high strength commercial	100 percent of allotment @ $4.14/hcf (allotment same as above) All other @ $4.37
Irrigation – Residential	Billed as if used through associated residential meter, OR annual allotment of 654 hcf/acre @ $4.14 (annualized allotments run July to June; new allotments available for the July water bill.) Unused allotments do not carry over. All other @ $4.37
Irrigation – Recreation/Parks/Schools	Annual allotment of 1,404 hcf/acre @ $1.95/hcf Next 240 hcf/acre/year @ $4.14 All other @ $4.37 (annualized allotment same as above)

At the time this rate structure was adopted the Santa Barbara City staff was gearing up for a drought and a comprehensive conservation program. IT staff had additional workload during the run-up to implementation but no new software was purchased. Utility staff reported improvement in long term water conservation and better management when rates are tied to discretionary needs of the customer.

CASE STUDY: MARCO ISLAND, FLORIDA

Marco Island is an upscale residential and resort destination community near the Everglades National Park. The water utility serves just over 8,000 residential connections and has limited potential for growth. The water utility was purchased from a private company in 2003. Water budget based rates were adopted in 2005 as a response to high customer usage and the high cost to expand the water storage, treatment and delivery systems. Moving from a flat fee rate system under private management, Marco Island adopted the schedule shown in Table 4.12 in 2006. Additional reasons for implementing water budget based rates included fairness on the part of the customers and a desire for better water management on the part of elected officials.

Table 4.12 Marco Island, Florida water rate structure and commodity charges

		Block 1		Block 2		Block 3	
Tier	**Lot Size (sq.ft.)**	**Gallons**	**$/kgal**	**Gallons**	**$/kgal**	**Gallons**	**$/kgal**
1	0 -9,000	21,000	$ 3.02	21,001 – 32,000	$ 4.53	32,001 +	$ 6.04
2	9,001 – 15,000	27,000	$ 3.02	27,001 – 41,000	$ 4.53	41,001 +	$ 6.04
3	15,001 – 20,000	35,000	$ 3.02	35,001 – 53,000	$ 4.53	53,001 +	$ 6.04
4	20,001 – 30,000	47,000	$ 3.02	47,001 – 71,000	$ 4.53	71,001 +	$ 6.04
5	30,001 +	73,000	$ 3.02	73,001 – 110,000	$ 4.53	110,001+	$ 6.04

Marco Island's city manager was the point of the spear, leading the change to the new rates. The idea for budget based rates originated with the hired consultant who crunched the numbers to assure the process was revenue neutral.

The political process went smoothly. A series of meetings was held to inform the residents and the water budgets were well advertised. A friendly local press helped present the rate system to the public. In the short time since implementation customer response has been "very positive."

Billing software was updated by the city's provider for less than $10,000 when it came time to change over to water budgets. It took three days for the software company to make the changes and train city personnel. No additional staff has been added because of the new rate structure.

There has been a lot of interaction with customers from the very beginning. When customers have concerns about their water bills, Collier County provides a free water audit service. Customer service representatives are empowered to provide credits to customers. Credits are allowed for new lawns, filling swimming pools, and leaks on the customers side of the meter if repairs are promptly made. The customer service representatives' decision can be appealed to the finance director.

CHAPTER 5

DESIGNING WATER BUDGETS

Two primary goals of any rate structure are to provide stable and sufficient income to the utility with some measure of fairness and equity. Water budgets purport to help with these goals while incorporating a third goal – improved water efficiency. In this chapter the researchers examine and explain different constructs for designing the water budgets for use with an increasing block rate structure. Some of the constructs discussed in this chapter have been implemented by agencies, as described in Chapter 4, while others are hypothetical examples of how water budgets could be developed by agencies in the future. The goal of this chapter is to present different concepts for the challenging task of constructing water budgets for the broad variety of municipal customers.

The central idea of the water budget system is to develop a reasonable measure of efficiency for each customer in a water utility. This requires specific data and information that will be different for residential, commercial, industrial, institutional, and municipal customers. As discussed in Chapters 3 and 4, a few utilities have implemented water budgets across all customer classes while others have focused on residential or irrigation-only accounts as a first step. This chapter is intended to add to the discussion of methods for designing water budgets.

OPTIONS FOR CONSTRUCTING WATER BUDGETS

Designing water budgets involves many steps, but two stand out as uniquely challenging. First, each community must decide on what it considers to be the accepted standard for the nature and intensity of water use by its citizens. Next these standards must be translated into specific volumes (water budgets or allocations) developed for each account.

Determination of the appropriate standards applies to each category of water use. Water budgets provide the opportunity to ask the question: How much water do we want set as a target for use in our community? For example, a community might decide to set residential budget allocations based on the expected water usage of older inefficient fixtures and appliances or based on the expected usage from high-efficiency fixtures and appliances. Or the community might start with budget allocations based on older fixtures (which would provide a larger indoor budget), but then over a period of time transition to lower indoor budgets based on high-efficiency fixtures and appliances. The approach to setting the budgets is a policy decision that needs to be made by the appropriate governing body, with public input.

Creating water budgets for landscapes involves a similar process. Some communities may wish to provide budgets that encourage a lush, well watered turf appearance, while others may wish to develop more restrictive budgets to encourage landscapes more appropriate for a drier climate. Clearly, the perceived surplus or shortage of the water resources will have a role to play as well as the historical practices, land use characteristics, and climate of the area. Landscape allocations for important public spaces like parks, civic centers, multi-family common areas, and playing fields may be set higher than those set for single-family landscapes. Allocations for vegetable gardens, swimming pools, hot tubs, and miscellaneous water features also need to be considered. If water budgets are to be perceived as "fair" it's important that they not be used in a punitive or heavy handed way which can be viewed as denying legitimate uses of water.

In setting budgets for non-residential (CII) customers a decision must be made about how best to develop the allocation. Two fundamental methods exist: (1) historical use patterns or (2) efficiency benchmarks in budget setting. In current practice, utilities that are creating budgets for CII customers use the historical patterns approach, although both IRWD and Boulder hope to improve upon this method in the future.

CII water use patterns are diverse, and it is difficult to determine how to establish an equitable budget that effectively rewards efficiency without being punitive. The historical usage pattern approach is an expedient starting point for CII budgets, but an unintended result can be to validate historical waste as part of the budget. Establishing efficiency benchmarks from the literature or from local field surveys offers an alternative approach, but requires data that are often not readily available.

Once policy decisions about the nature and intensity of water use and the methodology for developing budgets have been established, the process of actually setting the volumetric allocations for each customer becomes a more data driven and mathematical procedure. The goal of this procedure is to translate the water use policies into practical volumetric allotments of water for each customer in the system. The remainder of this chapter discusses possible techniques for budget setting in the various classes of customers.

Landscape Water Budgets

Every community has its own landscaping style and climate pattern. Fortunately the principles for determining the water requirements for lawns and gardens and trees are universal and can be applied in the effort to develop appropriate landscape water budgets.

The first step in creating landscape water budgets is to determine the appropriate reference evapotranspiration (ET) values to apply within the community. Reference ET (ET_o) refers to the evapotranspiration losses for a specific reference crop. In urban settings, 6" tall cool season grass in full sun and with a full water supply is normally used for determination of ET_o. The American Society of Civil Engineers (ASCE) has adopted the Penman-Monteith equation as the standard for calculating evapotranspiration (Jensen, et. al. 1990). In this method, values for ET_o are dependent on the temperature, solar radiation, wind and relative humidity. Note that precipitation is not figured into the calculation of ET_o. Areas with similar climatic conditions will have similar ET_o patterns, and a single community can have several distinct ET_o regions.

The next step involves deciding on a definition of the landscaped area to be included in the allocation for each customer. The water utility has little control over the landscaping decisions made by property owners so the definition of landscaped area must be broad enough to include potential changes made by residents. The easiest approach is simply to use the lot size or property area which is usually available from local tax assessor records. Since not all of a property is usually landscaped, this method can result in water budgets that provide more water than is required and hence do not convey an efficiency price signal effectively.

A more common approach is to use the concept of "irrigable area" which can be defined as: the total lot size minus the footprint of the house and an allowance for driveways and other paved areas. Using this definition, owners of similar properties in the same ET_o region would receive a similar landscape water budget. For some water utilities, this definition is similar to the definition of pervious area used for calculating stormwater fees. If pervious area data have already been calculated for each property this can significantly reduce the effort required to create landscape water budgets.

Using the "irrigable area" approach described above allows property owners to make their own decisions about how best to utilize the water provided in the allocation. Some owners may wish to include significant amounts of hardscape in their landscape design and then use water more heavily on the remainder of the areas, or perhaps have a pond or other water feature. The "irrigable area" approach is simple for the utility to administer and allows maximum flexibility in that it does not penalize owners (by reducing their budget allocation) for use of hardscape.

An alternative approach is to try and determine the actual irrigated areas of each lot using GIS or aerial photography. This usually involves identifying all areas on a property that are plant material and those that are not. This approach may lead to more accurate estimates of landscape areas, but as customers regularly make changes to their property the system must be updated to maintain accuracy thus increasing the complexity of program administration. Under this approach, property owners may find themselves with disincentives to remove turf for fear of having their water allocation reduced. Any of the approaches discussed here can work, but water agencies need to consider the advantages and disadvantages of each and discuss them with the community during the design of the water budget system.

Next, the type and relative areas of plant materials should be considered. These areas can be estimated for all lots based on percentage of maximum turf area, or individualized for each lot based on physical measurements. An easy approach that focuses attention of the water needs of different plants is to assume a limit on the percentage of the irrigable area planted in turf, and develop budget allocations based up this limit. For example, a utility might decide to budget enough water for turf on 50% of the landscaped area and provide sufficient water for less thirsty plants on the remaining 50%. Because the theoretical irrigation requirement depends so heavily on the plant materials used, the best way to tailor the irrigation requirement is to vary the areas of high and low water using plants.

ET_o is not a perfect measure of irrigation water requirements. There are several other factors which must be taken into account in order to arrive at that information. First, as noted above, ET_o does not include rainfall. To obtain a more accurate estimate of the actual water requirement of a landscape, local rainfall can be netted out of the equation to arrive at a net ET. ET_o is calculated for a uniform crop under ideal growing conditions and uniform water and solar exposure. Urban landscapes contain a myriad of different plant types, areas, micro-climates, slopes and exposures. Each of these will affect the water requirement for the landscape. ET_o is a measure of water requirement in the plant root zone. Since no irrigation system is 100% efficient in delivering water to the plant root zone some allowance needs to be made for unavoidable losses during application.

To arrive at reasonable landscape water budgets for the broad diversity of customers served by a utility, some sort of a "design standard" landscape form should be selected. At one extreme budget allocations could be established to provide a full supply of water for a 100% bluegrass landscape, with generous allowances for irrigation efficiencies. At the other extreme allocations could be set to provide sufficient water for a water-wiser (xeric) landscape. For many utilities the answer lies somewhere in between these extremes. Finding this happy medium involves deciding about the appropriate volume of water required to irrigate the "design standard" landscape.

Since most communities wish to set landscape budgets somewhere in between the two extremes outlined above, a sensible approach is to set budgets that are based on a limited amount of turf as a percentage of the total landscape. Allowances for irrigation efficiencies can be set based on well installed and maintained irrigation systems, thus encouraging efficient irrigation practices.

Detailed Landscape Budget Example

Table 5.1 shows a detailed approach to setting irrigation water budgets based on fundamental irrigation principals. The values shown in the table are for a hypothetical landscape containing 15,000 square feet of irrigated area. The areas for each of the irrigation zones are shown in Column 2. There are five plant types shown in the table: 1= cool season grass, 2= warm season grass, 3= combined grasses, 4= moderate water using plants, 5= low water use plants.[15] Plant type information is shown in columns 3 and 4. Columns 5 through 8 represent various coefficients used for determining the proper water budget. The density coefficient (K_d), shown in column 5, provides a measure of the percent of the ground that is covered by the plant material in question. The microclimate coefficient (K_m), shown in column 6 measures the relative exposure of the zone to sunlight. The crop coefficient (K_c) is a measure of the percent of water that the individual plant material would use compared to the reference crop, all other factors being equal. The zone coefficient (K_z) is the product of the other three coefficients.

The plant water requirements for each zone are shown in Column 9. These are calculated as the product of the annual ET_o times the zone coefficient times 0.623, which converts from inches to gallons per square foot. The ET_o values used for this site are shown in Table 5.2. ET_o data are normally presented in inches of water equivalent. Inches can be converted to gallons per square feet by multiplying by 0.623 - the ratio of 7.48 gal per cubic foot/12 inches per foot. This is far more convenient for calculating water budgets since it can be multiplied directly by an area to arrive at a volume of water required by the zone during either a month or a year depending on the interval under consideration. In Table 5.1 the annual ET value is multiplied by the area in order to determine the number of gallons per year required by each zone. The final calculation is to divide the irrigation water requirement for each zone by the assumed efficiency of the irrigation system, shown in Column (10). This provides the theoretical irrigation requirement for each zone. The total theoretical irrigation requirement for the landscape is shown as the sum of the individual zones. In this case the landscape would require 248,211 gallons per year to meet its theoretical irrigation water requirements.

The last two rows of Table 5.1 compare this landscape to a hypothetical landscape planted in the reference crop (6" high cool season grass). This reference landscape (i.e. the same property planted exclusively with cool season grass) would require 325,118 gallons/year to meet its irrigation requirement. The ratio of the actual irrigation water requirement for the landscape to the reference water requirement is a useful number, called the landscape coefficient (K_l). This coefficient makes it simple to compare the water use of the landscape to the irrigation demand based on ET_o, which is a widely available reference number. In this case we know that the landscape coefficient is 76%, so this property is expected to have a water demand of 76% of the

[15] This example uses five different plant types, but the results would be equally valid if only turf and non-turf plant types were used. The key is that the calculations can be a detailed or general as appropriate to the needs of the community. General estimates can be used for established areas, and more detailed calculations can be done for new landscapes on a case by case basis.

ET_o based demand. This is a value that could be generalized for other similar landscapes and for preparation of water budgets.

Other factors that need to be considered in setting landscape budgets include the use of historic or real-time ET. San Juan Capistrano and IRWD are the only agencies currently using real-time ET to set water budgets. Basing budgets on historic ET is simpler, but less flexible. If a single estimate of ET_o is used, should it be the average or dry year value? If real time ET is used to calculate budgets, the customer won't get his exact monthly water budget until *after* the billing period is over. Although this sounds confusing it has not been problematic for San Juan Capistrano or IRWD. They provide an estimate of the next month's water budget on each bill with the caveat that it could be adjusted up or down depending upon climate conditions. Another factor is rainfall, which can be deducted from the irrigation water requirement (real-time) if it is significant in the area.

The example shown in Table 5.1 requires far more detailed landscape data than most water agencies maintain or have access to. How can these data be obtained? New customers could be asked to submit landscape information within a set time frame as a condition of service. Data for existing customers could be obtained from user surveys or GIS data. An alternative approach is to conduct detailed analyses of the landscapes from a substantial random sample of lots in the community and to develop curves of irrigated area and percent turf versus total lot area for the sample. These curves could serve as default values for the existing customers. If additional accuracy is required, customers could update the default landscape values with their own data, subject to verification by the utility.

When landscape budgets are established a decision needs to be made as to whether the budgets will be set on an annual, quarterly, bi-monthly, or monthly basis. Most water agencies use their billing cycle duration (usually monthly but occasionally bi-monthly) as the time increment for water budgets. If fixed monthly budgets are used then consideration should be given to balancing out normal variations in monthly demands from weather. One approach is to set the monthly demands using dry year conditions. This will almost always provide adequate (or more than adequate) budgets, but won't encourage conservation. Another approach is to allow for the carryover of "saved" water in one month for use in latter months within the year. No current water budget system has a carryover provision, but the Otay penalty budget system did. The third approach is to use real time weather data for determining water budgets.

A final consideration in setting landscape water budgets is the linkage between the water budgets and the drought response plan. This issue is taken up in detail in Chapter 6. Most landscapes can tolerate some dormancy period without significant loss of plant material. Some plants are hardier than others and some cool season turf grasses are known for their ability to revive after drought periods. The key is in designing a curtailment program in advance of the drought that will generate the most saved water with the least permanent damage to the landscapes. Such plans could start by reducing irrigation of turf areas in a way that encourages them to go dormant. Irrigation and mulching plans for deep rooted trees and shrubs would be developed to carry these over with infrequent but deep watering.

Table 5.1 Detailed Approach for Determining Landscape Water Budgets

Zone (1)	Irrigated Area (2) (sf)	Plant Code (3)	Plant Type (4)	K_d (5) (density)	K_m (6) (climate)	K_c (7) (crop)	K_z (8) (zone)	Plant Water Requirement (V_p) (9) (gpsf/yr)	(gal/yr)	Zone Efficiency (10)	Theoretical Irrigation Requirement (11) (gal/yr)
1	1000	1	Cool Season Grass	1.00	1.00	0.95	0.95	20.6	20,595	80%	25,744
2	2500	2	Warm Season Grass	1.00	1.25	0.65	0.81	17.6	44,036	80%	55,045
3	3000	4	Mixed Water Using Plants	0.75	1.00	0.63	0.47	10.2	30,730	90%	34,145
4	3500	5	Low Water Using Plants	0.80	0.85	0.45	0.31	6.6	23,218	65%	35,721
5	5000	3	Combined Grasses	0.90	1.00	0.75	0.68	14.6	73,167	75%	97,557
Total	15,000						Theoretical Irrigation Requirement (gal/yr)				248,211
								Reference Application (gal/yr)			325,118
								Landscape Coefficient (K_l)			76%

K_d = Crop density coefficient, the percent of ground that is covered by the plant material in question
K_m = Micro-climate coefficient, the relative exposure of the zone to sunlight
K_c = Crop coefficient, percent of water that the individual plant material would use compared to the reference crop, all other factors being equal
K_z = Zone coefficient = $K_d \times K_m \times K_c$
K_l = Landscape coefficient, ratio of the actual irrigation water requirement for the landscape to the reference water requirement

Table 5.2 Monthly and annual ET_o

Month	**ET_o (inches)**
January	0.16
February	0.30
March	0.56
April	1.63
May	3.52
June	6.58
July	8.63
August	6.67
September	4.04
October	1.87
November	0.64
December	0.20
Total (inches)	**34.80**
Total (gpsf)	**21.58**

Note: 1 inch of water = 0.62 gpsf

INDOOR WATER BUDGETS

This section examines methods for establishing indoor water budgets. The most important factors in determining indoor budgets for residential customers are the number of occupants in the dwelling, the size of the dwelling, and the types (and efficiencies) of water using fixtures and appliances present in the home (DeOreo, et. al. 2004). This naturally leads to ways to establish indoor budgets for residential customers. Setting indoor budgets for non-residential customers is more complicated, but there are some principals that can be followed to help make the process reasonable.

Residential – Single-Family

The simplest method of setting indoor water budgets for residential customers is to base them on historical per capita use. This implies that the indoor budgets will assume whatever level of efficiency was current in the system at the time they were adopted. In such a system, if average single family indoor demands were 5500 gal/month with an average of 2.7 persons per dwelling then the indoor budgets would be set at 67 gcd. This is a demand that is typical of homes with older (pre-Energy Policy Act, 1992) technology[16]. Using this approach, indoor budgets could be set at a default of 5500 gallons/month with provisions for adjustments based on more occupants, or special medical needs. In the water utilities researched for this project most set their single-family indoor budgets at a fixed amount ranging from 5 to 7 kgal per month. Only IRWD uses a daily per capita value of 75 gcd to develop indoor water budgets for single-

[16] Energy Policy Act (EPAct) of 1992 specified maximum flow rates for showers at 2.5 gpm and toilets at 1.6 gpf.

family customers.[17] A family of four in IRWD would receive a monthly budget of approximately 9 kgal per month.

If indoor water budgets are set assuming standard (pre 1992 EPAct) technology then customers who have invested in water conserving fixtures and appliances would normally use less than their budgeted amount. Consideration should be given to how this investment should be rewarded. One obvious way would be to establish a water conservation rate tier that is set lower than the standard rate for water used within the customer budget. A rational method to set this type of rate would be to allow customers investing an amount deemed necessary to pay the incremental costs for high efficiency devices to recoup this investment over a five year period. Both IRWD and Boulder use their first tier as an encouragement to conserve.

A more sophisticated manner of setting water budgets for single-family indoor use recognizes that indoor demands do not increase in direct proportion to the number of people living in the home, but the additional increment of demand is smaller for succeeding occupants. This is shown in Figure 5.1, which is based on detailed end use studies in Seattle, WA, Tampa, FL and the East Bay Municipal Utility District, in Oakland, CA (Aquacraft, 2005). This figure, adapted specifically for this study, shows the measured indoor water use for approximately 100 homes before and after they were retrofit with water conserving fixtures and appliances. The top line shows that the indoor water use for the houses in their original condition (circa 2002) followed the relation: $y=87.4\ x^{0.687}$, which is significantly different from the linear case. A house with 1 single occupant was found to use approximately 88 gallons per day (gpd) (88 gallons per capita per day (gcd)) for total indoor household use. Doubling this to 2 occupants was found to raise the water use to approximately 140 gpd (70 gcd), rather than 196 gpd predicted by a linear approach. An "average" home with 2.7 occupants would use approximately 173 gcd (64 gcd) for indoor uses. Demands for homes with different numbers of occupants can be derived using the formula.

Using relationships derived from detailed measurements such as the previous example provides an empirical basis for making adjustments to indoor water budgets rather than having to rely on either overly generous allocations or arbitrary values.

The second (diamond) line in Figure 5.1 shows the water use in the homes after high-efficiency toilets, clothes washers, showerheads and faucet aerators were installed. The equation describing household water use for these high-efficiency homes changed to: $y = 50.2\ x^{0.767}$ gpd, which is also a non-linear relationship. Using this relationship, homes that are similar to the standard homes in size and the number of persons, but with water conserving toilets, showerheads, faucets and clothes washers would use only 107 gpd (39.9 gcd) for indoor uses.

There should be considerable latitude for water agencies in setting water budgets for indoor single-family demands. The factors that affect indoor water use are well understood and established with empirical data. Budgets can be set using relatively simple parameters and still work quite well. As more precise data become available it can be used to refine the budgets.

[17] IRWD is currently reviewing this gcd value. The 75 gcd was established in 1991 prior to the passage of the 1992 EPAct.

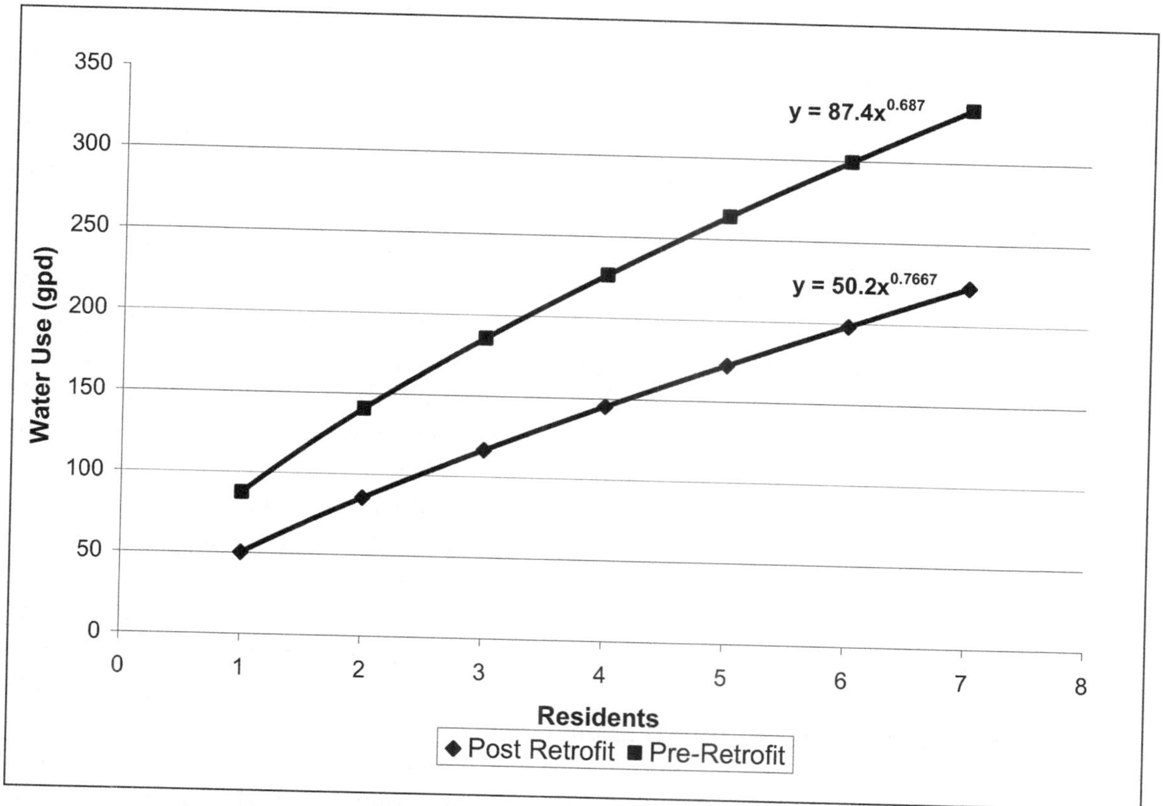

Figure 5.1 Household water use vs. number of residents

Residential – Multi-Family

Setting indoor use budgets for residential multi-family accounts requires a minimum amount of information about each account. The average per unit indoor water use for apartments in the United States was measured to be 52 kgal/year (Mayer, et. al. 2004). This is nearly 20,000 gallons lower than indoor use for single-family houses, which averages approximately 71 kgal/year (Mayer, et. al. 1999). Key factors determining water use in multi-family housing include (Mayer, et. al 2004):

- Whether the project was built prior to 1995
- If the project is a rental, versus a condominium, etc
- If the units are individually sub-metered
- The average commodity charge for water and wastewater
- The number of bedrooms per unit

This illustrates how a utility could customize multi-family indoor allocations through surveys and data analysis to explore the relationships between water use and key contributing factors. In practice, most multi-family indoor water budgets are based on a fixed per housing unit volume.

Commercial and Institutional Customers

Developing water budgets for commercial, institutional, and industrial customers (CII) probably poses the most difficult challenge in the allocation process. The CII sector is diverse and multi-faceted and uses water for many different purposes with a wide variety of equipment. Only a handful of agencies have included CII customers in their water budget rate structure program. Water agency approaches to developing CII budgets usually include relying on historic usage patterns to establish a baseline budget. The problem with this approach is that it rewards customers whose use is historically inefficient with a water budget that "grandfathers" in that in-efficiency. The alternative to this approach involves developing a better understanding of how water is used in the CII sectors and basing water budgets on efficiency benchmarks for classes of CII customers. This approach is examined in detail below.

To begin an analysis of CII water uses and budget allocation methods it first makes sense to divide the group into two categories. Commercial and institutional users (CI) tend to be more similar from system to system with more consistent demand patterns. Industrial customers are far more diverse and frequently obtain large portions of their water from private supplies. In the water budgeting process industrial customers should be evaluated on a case by case basis so that fair and justifiable budgets can be developed. In the next section of this report the authors discuss commercial and institutional users (CI) as one group and industrial users separately.

At first glance the complexities in setting CI water budgets may appear insurmountable. The problem, however, becomes more manageable when one considers that after irrigation is eliminated nearly 70% of all CI water use occurs in just 10 categories of CI customers (Dziegielewski et. al. 2000). Figure 5.2, which was prepared with data from the AwwaRF Commercial and Institutional End Uses of Water study (2000), illustrates this point. These data were obtained from five large water agencies in the southwestern U.S.: Los Angeles, Irvine Ranch, San Diego, Santa Monica and Phoenix. Other regions may vary, but given the nature of commercial and institutional uses these values should be a reasonable indicator of consumption patterns in the CI sector. CI water demand and end uses are an important area for future research.

In order to set budgets for CI customers the first step is to disaggregate the CI customer class into appropriate sub-categories within the utility billing system such as office buildings, restaurants, schools, hotels, car washes, etc. Since each sub-category will have its own end-use patterns it is logical to compare use within each sub-category rather than attempting to compare water use in schools to restaurants or hotels.

To make the system more responsive the categories shown in Figure 5.2 could be usefully sub-divided even further. The schools category could be broken down into primary, secondary and colleges; hotels could be divided into hotels, motels; laundries could be divided into self service and full service laundries, etc. The North American Industry Classification System (NAICS) which has replaced the U.S. Standard Industrial Classification (SIC) system offers an excellent solution for water agencies seeking a standard classification system for CI customers.

A fundamental point when making comparisons of CI water use is to examine as similar a set of customers as possible. Since the 2000 AwwaRF study offers the best available data on CI end uses, working with the ten categories shown in Figure 5.2, is a reasonable starting point.

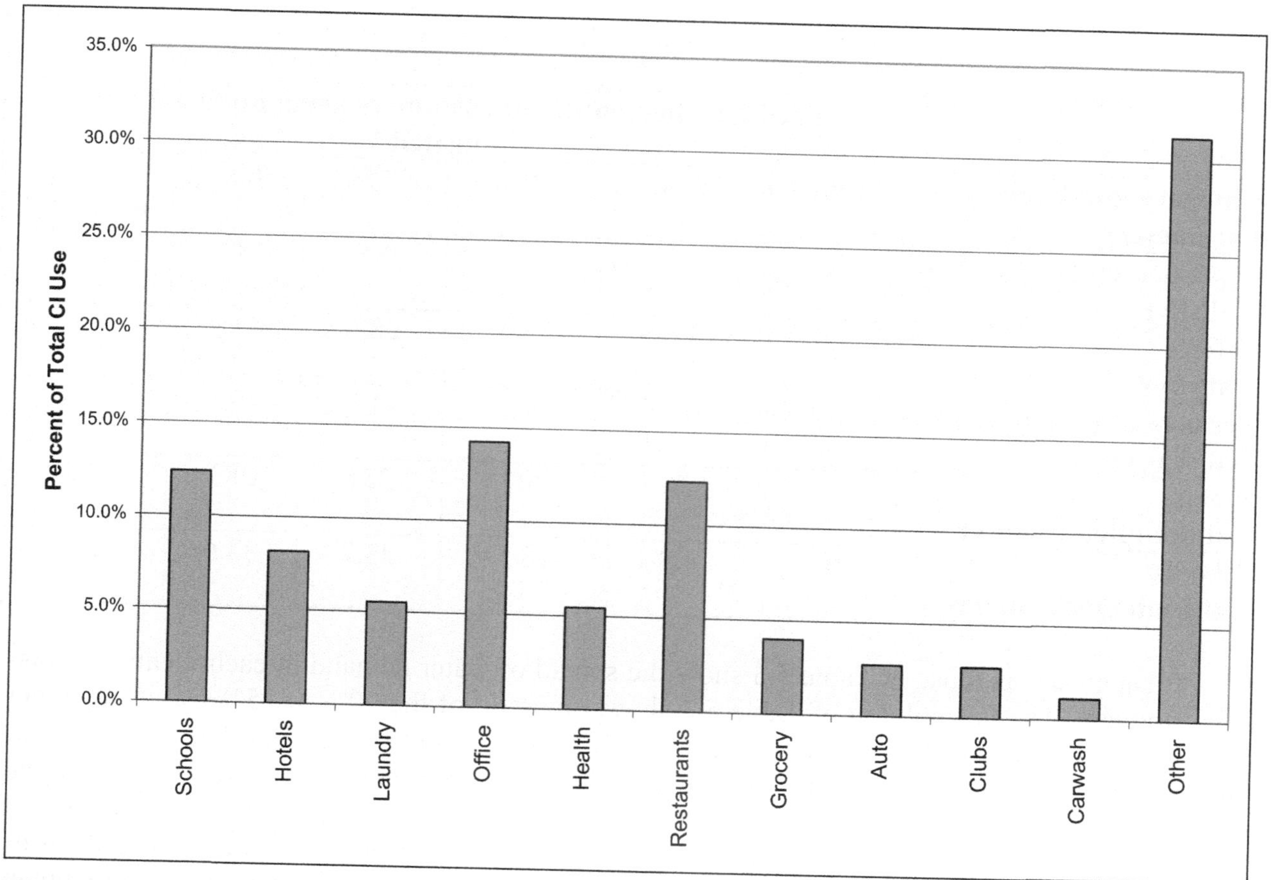

Figure 5.2 Water use by 10 important CI categories

Once customers have been grouped into logical sub-categories, water use can be examined for each sub-category. Then it becomes possible to begin the process of setting indoor budget allocations using efficiency benchmarks that apply within the groups. The use of CI benchmarks for setting indoor water budgets appears to be an important step forward from use of historical data, which may or may not reflect efficient use.

In the AwwaRF Commercial and Institutional End Uses of Water study efficiency benchmarks were developed for the five most important sub-categories of CI water use (Dziegielewski et. al. 2000). Water use for each sub-category was normalized using an appropriate and readily obtainable variable. Distributions of the normalized water use data were calculated from large groups of similar customers and patterns observed. These data are shown in Table 5.3

With the type of data shown in Table 5.3 it becomes possible to set water budgets for CI customer in scientifically rigorous manner designed to reward and encourage water efficiency. These data tell a utility what percentage of its customers in each sub-category are likely to fall under or over a particular usage rate, and by what percentage. This will allow the water agency to understand the impacts of a given budget system in terms of how many customers might fall into the high cost blocks. Setting budgets based on an efficiency benchmark prevents allocations from being set in a manner that rewards inefficiency.

Table 5.3 Efficiency benchmarks for five CI categories

	Predicted percentiles of customers using up to amount indicated					
Category and Usage Parameter	**10%**	**25%**	**50%**	**75%**	**90%**	**Ratio 90%:25%**
Supermarkets (w/cooling) **Gal/sf/yr**	46	52	59	85	89	1.7
Offices **Gal/sf/yr**	9.5	13.6	21.9	28.6	41.9	4.4
Restaurants (full service) **Gal/seat/yr**	26.1	30.5	34.7	41.9	49.7	1.6
Hotels **Gal/occupied room/yr**	107	136	207	251	283	2.1
Schools **Gal/student/school day**	10.3	14.8	18.5	23.7	33.6	2.3

Data such as those in Table 5.3 show the spread of water demand in each identified class of customers. The last column of Table 5.3 shows the ratio of the 90% to 25% use. The larger this ratio, the greater the variability of water usage in the group. It turns out that for this set of customers the greatest variability is in office use, which has a ratio of 4.4:1. The groups with the most consistent water use patterns were supermarkets and full service restaurants. Hotels and schools fell in between. These data can assist a water agency in predicting which sub-categories of customers can be fit with a single parameter water budget (i.e. supermarkets and restaurants in this case). The data also suggest that because of the diversity of demand within the sub-category, office buildings will probably need a more custom analysis or further sub-categorization.

One option might be for a water utility to use the 25 percentile unit consumption as an efficiency benchmark for restaurants and supermarkets, the 50 percentile for hotels and schools and the 75% for office buildings (or another combination of values that produces desired results). Access to the percentile consumption values for CI sub-categories could provide a water agency setting up water budgets a rational basis for choosing the appropriate levels for each sub-category of customers. This appears to be a promising (although data and labor intensive) method for establishing reasonable and conservation-oriented CI water budgets. This approach is another topic for future research.

Industrial Customers

As noted above, industrial customers are far more diverse even than CI customers. Furthermore large industrial users frequently obtain large portions of their water from private supplies. In the water budgeting process, industrial customers should be evaluated on a case by case basis so that fair and justifiable budgets can be developed.

In setting water budgets for industrial users using the efficiency benchmarking approach discussed in this chapter, the process should begin with a thorough survey of the industrial customers supplied by the agency. How much water do they currently use? What trends for their demand can be found? Is consumption per account increasing or decreasing? Surveys should be conducted that will allow the agency to link water use to production of finished products or dollars of output. Industries that have a more intensive use of water per unit of output, and which consume larger volumes of water from the system will be the industries of most significance from the perspective of water conservation. These industries will also be the most vulnerable to shortages in water supply which are apt to lead to serious economic dislocations within the community during drought.

Industry has two primary motivations for conserving water: (1) to reduce costs; and (2) to reduce risks of interruptions to production (Wade, et. al. 1991). These two concerns often lead industries to undertake serious water conservation efforts on their own; in the interests of their own self preservation. In many cases the activities of industries have focused on the least cost and easiest measures for conservation. Additional conservation may be possible, but at higher costs. The water agencies need to obtain data from actual industries in their service area in order to assess the level of conservation and potential for additional conservation.

Industrial customers have three key concerns regarding water reliability (Wade, 1991). First, as waste is eliminated any shortages in water supply will be more likely to lead to interruptions in production. This issue is often referred to as demand hardening. Second, industry has paid a premium for saved water over costs from the water agency as a price for increased reliability. Third, industries would prefer to have a set allocation of water that was assured from the agencies (i.e. a water budget) at costs lower than they are paying for internally conserved water (Wade, 1991).

If these conclusions are correct then many industries have a major incentive to work with water agencies in developing water budgets that are based on reasonable levels of conservation and elimination of waste within their operations. These budgets could be set on a case-by-case basis. Setting industrial budgets in this manner addresses the three concerns discussed above. Since the budgets would be set knowing that the best available technologies are in use for water conservation they would not necessarily be reduced as a simple percentage of use in drought times. Specific drought operations could be identified for reducing water use in times of shortage, but the system could be tailored to manage shortages in a controlled manner with the minimum of disruption to production. This would allow industry to determine which internal conservation measures to undertake and how much to spend on them based on an assurance from the utility of how much water can reliably be supplied.

A few categories of industrial users consume the bulk of all industrial potable water. Seven sub-categories (food, refining, chemical, and concrete, aircraft, paper and computer/electronics) accounted for over 92% of all potable water use by industrial customers (Wade, 1991). These data were obtained from California, and other areas may have their own unique industries (such as steel making or aluminum production). Nonetheless, the principle of identifying the locally relevant industries is a generally applicable first step in the water budgeting process.

The next step in establishing industrial water budgets is assessing the potential for water conservation projects within the industrial users served by the agency. These can be exhaustive studies or based on samples. The most accessible conservation opportunities should not be included in the budgeted amounts. These include things such as leak repair, use of inefficient sanitary fixtures and appliances, once through cooling and once through process uses that could easily be replaced by recirculation. Industries may be willing to undertake all of these conservation programs if they can be assured that the amount allocated for their use in their budget is a firm number that will only be curtailed in severe shortages.

After the current state of conservation in the industrial customers is understood the potential for additional conservation should be explored. It is likely that this water will be much more expensive than the water in the initial group of conservation work. Knowing the cost for the additional conservation will help in determining how to establish drought targets.

In addition to knowing the costs for additional conservation, there should be an understanding of the damages that are likely to occur from water shortages to the various industries. Industries that implement basic water conservation programs and are using good technology to eliminate waste in their facilities need to have a water budget that reflects these investments. Conversely, industries that employ many wasteful or inefficient uses and processes should not have these included in their budgets.

CONCLUSIONS

Water budgets should be objective and must make sense to customers. The central idea of the water budget system is to develop a reasonable measure of efficiency for each customer in a water utility. This requires specific data and information that will be different for residential, commercial, industrial, institutional, and municipal customers.

Designing water budgets involves many steps, but two stand out as uniquely challenging. First, each community must decide on what it considers to be the accepted standard for the nature and intensity of water use by its citizens. Next these standards must be translated into specific volumes (water budgets or allocations) developed for each account.

To the extent possible water budgets should be based on empirical data (indoor budgets) and horticultural science (outdoor budgets). Be explicit about the formulation of the allocations and be up front about areas where agency discretion enters into the allocation formula.

CHAPTER 6

WATER BUDGETS AND DROUGHT RESPONSE

"When the wells dry, we know the worth of water."
– *Benjamin Franklin, Poor Richard's Almanac.*

Water budgets and water budget rate structures offer water utilities powerful new tools for reducing demand during drought and for monitoring customer compliance with drought restrictions. This chapter examines the potential of water budgets as a drought response tool and will examine the benefits and challenges of this method of drought response. A case study from the East Bay Municipal Utility District (EBMUD) that implemented a water budget system as a drought response tool is also presented in this chapter.

Historically drought has been considered an unpredictable natural disaster, and in response to drought water system managers have typically relied on predictable response mechanisms. Water managers have viewed droughts as low water supply events of unknowable and uncontrollable duration. In response water utilities typically invoke some form of water use restrictions on customers that limit outdoor use, often using somewhat arbitrary baselines such as the previous year's use. The strategy for drought management has been expanded in recent years to put more emphasis on drought planning instead of mere drought response. The emphasis has changed to a more pro-active rather than reactive mode of thought (Billingsley, 2004).

Drought management plans may vary from place to place, but progressive drought plans generally include four main elements (Lampe, et. al., 2003). Each of these four elements contains sub-elements that must be taken into account in order to formulate a successful drought response plan.

1. An analysis of how water supplies are likely to vary for a range of droughts;
2. An understanding of local water requirements and the ability to reduce them;
3. Establishing measures for reducing demands in an equitable manner;
4. Creating a modern system designed for flexibility to match the conditions of the drought.

A water budget rate structure is a sword that cuts both ways during drought. First it establishes an empirical and quantifiable limit to the amount of water that a customer is entitled to use at a given price from a given tap. Second, it theoretically reserves a volume of water that is set aside for the customer to use as he sees fit. Water budgets have the potential to protect the utility from overuse and to protect the customer from having her water allocated to other uses or micromanaged by the utility. In time of shortages, water budgets allow a water provider to quickly and easily identify excess use and even penalize it if necessary. By summing all water budgets, utilities can quickly understand the amount of water likely to be required to meet customer demands in any given month. During a drought, water budgets have the potential to assist water utilities in more fairly apportioning demand reductions among customers with different needs and among different customer classes since the reference point for reductions is based on the water required by each customer in normal times. Historically, when customers are asked to reduce their use from the previous year, justified complaints arise from customers who are already conserving, and don't have as much room for additional curtailments.

MOTIVATING DROUGHT PLANNING

When droughts occur infrequently there is an unfortunate (and perfectly normal) tendency to neglect drought preparation and planning until the last minute while staring down the barrel of the next drought. During periods of adequate supply, utilities often aren't motivated to spend the effort and money needed to develop a comprehensive drought response plan. One of the most pressing needs in improved drought planning is raising the awareness of the nature of drought in the minds of the planners and customers when there is no immediate crisis.

One way to raise the awareness of droughts during times of sufficient supply is to make people change the way they think about droughts. Droughts should not be thought of as unpredictable disasters that come out of thin air, but rather understood as part of the predictable natural pattern of hydrology. In many regions, every year has the risk of being a drought year. This raises the importance of a drought plan with an early warning system. This is similar to the concept of flood risk, and just as there is a 100 year flood to consider there is also a 100 year drought to consider. Every water system faces a certain risk of drought. This knowledge places responsibility for drought planning on the local community, as opposed to seeking emergency relief from drought from state and federal governments on the basis that the drought could not have been anticipated.

DROUGHT PLANNING AND WATER BUDGET SYNERGIES

Water budgets can enhance drought planning in a number of ways. The process of setting up water budgets provides a good foundation for drought response. The National Drought Mitigation Center has developed a ten point description of the drought planning process (Wilhite 2005). As a nationally recognized leader in drought preparedness, these steps offer a broadly accepted model for drought planning. In a number of instances these drought planning steps parallel the process used in developing a water budget rate structure. In addition, the information generated in the water budgeting process can be useful in developing, planning, and executing a drought plan. Developing both a water budget rate structure and a drought response plan can improve water management more than either effort on its own. The synergies between the steps of the drought management planning process are particularly aligned with a water budget rate structure and all that goes into its development.

Step 1: Appoint a Drought Task Force

The first step in drought planning is to appoint a panel to oversee the formulation and implementation of the plan. A body of this kind is essential to the process because it improves the likelihood of community buy-in to the plan by promoting involvement by the local stakeholders. A drought task force typically includes a broad spectrum of experts and stakeholders in a community including civic leaders, business leaders, the landscape ("green") industry, homeowners, utility personnel, water resources professionals, and others. A similar task force (or the same body) can be helpful when developing a water budget rate structure. A key linkage between an agency's drought response and their water budgets must be established. The East Bay Municipal Utility District (EBMUD) case study (presented later in this chapter) provides an illustration of the uses of water budgets to fairly apportion drought usage reductions and to effectively penalize customers for non-compliance with restrictions.

Step 2: State Purposes of Drought Plan

A broad statement of purpose for a drought plan is: To lessen the impacts of drought by developing a set of mitigating activities and programs that change basic vulnerabilities. A drought plan is intended to provide government with tools and means for assessing drought conditions, creating mitigating programs and actions prior to a drought, and establishing response options that reduce economic damage, environmental losses, and social hardships during drought (Wilhite et. al. 2005). Curtailing water use is an essential part of the drought response process, and the size of the curtailments plays a major role in determining customer vulnerabilities and potential economic damage. This step is closely linked to the water budgets since the goal of establishing water budgets is to identify the legitimate needs of all customers for water. Water budgets are defined specifically to identify the legitimate water needs of the customers and efficient levels of usage versus waste. Development of water budgets are thus an excellent way to understand the baseline demands in the system and the amount of curtailment that is possible, both from reductions in waste and in normal use by the customers.

Another goal of drought planning that is assisted by the water budget process is that of treating all customers fairly during a drought. In this context, fair treatment implies two things: (1) All uses of water are examined for drought curtailments, rather than just a few obvious uses such as landscape irrigation; and (2) the economic losses during the drought are apportioned as equitably as possible among customers. In the traditional response mode of drought management, watering restrictions are often used to curtail large and obvious uses of water such as landscape irrigation. This leaves many potential areas for water savings untapped and does not fairly apportion usage reductions among all customers.

When all customers have water budgets, they can also be provided drought usage targets based on the water budget to set curtailment goals when necessary. Water budget-based curtailment goals have the potential to involve the entire customer base in the drought response effort while providing individual customers with flexibility and choice in meeting their goal. From the water utility perspective, the budget-based curtailment goals offer a regular and accurate method to monitor compliance with drought restrictions, implement drought pricing, and even enforce drought reductions if necessary. The development of water budgets thus can create opportunities for drought response and enforcement previously unavailable to water utilities and customers.

Step 3: Seek Stakeholder Participation

A water budget rate structure can be an effective method for encouraging stakeholder participation in the drought preparation and response process. Implementation of a drought plan only has an impact on water users during the drought events. Water budgets, however, can impact water use ahead of a drought. The process of creating water budgets typically involves stakeholder involvement as well as more extensive utility outreach to insure that all water users are aware of how the system works and will impact them. When using water budgets as an element of drought preparation, stakeholders may be invited to review proposed drought budget allocations and the reasonableness of proposed drought pricing and enforcement measures. The process of setting up water budgets can encourage customers to contemplate their water use patterns, determine the extent to which they can be made more efficient, and plan for curtailment during drought.

Step 4: Inventory Resources and Identify Groups at Risk

This drought preparedness step illustrates the close relationship between establishing water budgets and managing overall water resources for a community. The development of water budgets can augment the process of defining the available water resources and can identify customers at greatest risk from drought.

Droughts are manifested in the shortages they create to utilities and most importantly, water users. Understanding the variability and reliability of the water supply for droughts of various frequencies and severities is essential. If curtailment is required, how will it be implemented? Should curtailments be based on a uniform reduction from established water budgets, or should consideration be given to the potential for differential damage suffered by distinct customer classes? For example, an efficiently run commercial operation with limited outdoor use might suffer severe economic damage from a level of drought curtailment that might be easily accomplished by a typical single family residence. Identifying these "at drought risk" customers is made easier by the water budgeting process. Once identified it may be possible to create customized drought reduction targets that more fairly apportion the economic damages of a drought.

From an administrative perspective the simplest method of allocating shortages is through a reduction in all outdoor water budgets by the same fixed percentage. This approach has the best chance of success when dealing with customers in the same category. For example, if all single-family water budgets are set up using the same set of rules, then making across the board reductions as a percentage of the budgets should work. When working with customers in different categories, however, consideration should be given to the relative elasticity of the water use by each. As shown in Chapter 2, outdoor use is far more elastic than indoor use. It is also considered more "discretionary", although serious gardeners might dispute this point.

The problem of establishing fair curtailments among dissimilar customer categories is demonstrated in Figure 6.1, which shows curves of how the percent of lost value from use of water increases as a category of customers' water supply is curtailed. In this hypothetical case customer category B is far more risk prone to drought than is customer category A since for any percentage of water use curtailment category B will suffer far greater economic losses than will category A. We are assuming that both categories have properly established water budgets and that the percent shortages are relative to those budgets. This figure highlights the problem of setting equal curtailments for all customers (as a percentage of their budgets). In our hypothetical case, a 50% curtailment would result in an economic loss of 18% for category B, while category A would suffer only a 5% loss.

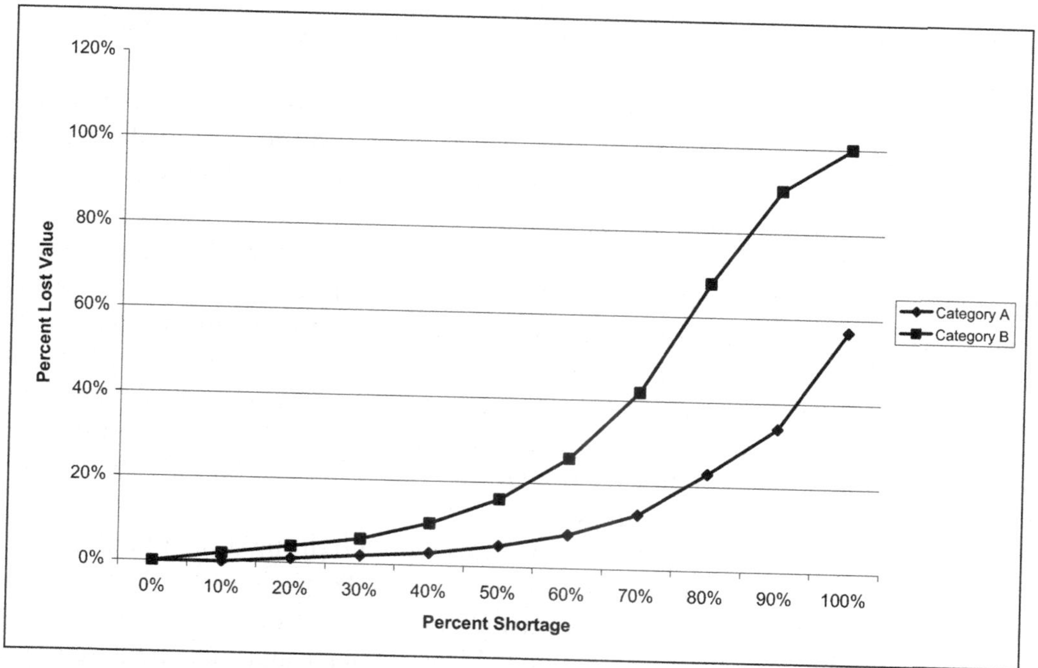

Figure 6.1 Identification of At-Risk customers, hypothetical shortage vs. lost value curves

If a situation similar to the example in Figure 6.1 occurred, and if it is assumed that both customers are already using water efficiently then arguably the fair curtailment approach would be to ask Customer B for a 20% reduction in use and Customer A for a 50% reduction. This would result in both customers suffering an equal loss of around 5% of the value of their water use.

Water budgets may assist in dealing with the differences in drought risk. As water budgets are developed for different customer categories, those most vulnerable to shortages may become evident. In the example above, vulnerability to a shortage is equated to suffering economic losses for even the first increment of shortages. Customers that suffer large losses with even small shortages of water can be classified in economic terms as having inelastic demands. The water budgeting process can help identify these customers so that some thought can be given to proper drought response. It may be that there is simply no way to reasonably reduce the water use, and that the value is so high (e.g. a hospital or dialysis unit) that the supply must be protected by the community. On the other hand, process changes or new technologies may be available that can increase the flexibility of use and the ability to curtail use in drought times. Typically these changes require time and preparation that are not available during a drought. But time could be available during the water budget planning process. In the right circumstances, the development of water budgets can amplify the process of first identifying customers at risk, and second, in increasing their flexibility of use prior to drought onset.

Step 5: Prepare and Implement Drought Plan

Water budgets assist with drought plan preparation because almost all of the work done in establishing the water budgets is also needed to prepare the drought plan. During the water budgeting process, if consideration is given to the issues of fair allocation during drought and appropriate curtailment approaches, then much of this effort can be used as part of the drought response plan. In this sense, most water budget rate structures form the basis for a drought response plan.

A water budget rate structure also provides an effective drought enforcement mechanism. It provides the utility with more certainty that the drought restrictions will be complied with, while at the same time providing the customers with the maximum flexibility to use whatever water they are allocated when and how they choose.

Water budget rate structures can help with drought plan enforcement in the area of communications. The water budget rate structure, with its billing system, informs all customers on a regular basis of the required use reductions. The water bill can show each customer how much water she is allocated during the drought. This information can theoretically be developed well before the drought occurs as part of the budgeting process. This is a far more reliable and effective way to implement drought-related conservation since it is pre-planned rather than improvised. The billing system is already in place and the bills can provide the public with the information needed to respond to the drought.

Another way that water budget rate structures aid with drought plan implementation is in the enforcement of mandatory demand curtailment. A simple query of the billing database can inform the utility each billing period which customers have complied with drought restrictions and remained within budget and which have not. If the higher water rates being charged are not sufficient to elicit co-operation then additional fines and penalties can be considered. This is a highly reliable system. Unlike the "water cop" approach where customers are ticketed if they happen to be observed violating the drought restrictions, a water budget drought enforcement program automatically identifies every customer who is not complying, thus enabling fair and uniform enforcement. Water enforcement patrols are costly and can only catch violators "in the act" of violating a watering restriction. A water budget, however, provides a regular and automatic check on which customers are in or out of compliance with drought response.

Enforcement of drought restrictions using the water budget approach is far simpler for the water utility and the customers than the "water cop" approach. If the curtailments have been planned and designed well ahead of the drought there should be less discussion and debate prior to implementing the curtailment program. If the onset of drought is rapid, the utility is prepared for rapid response. Once the curtailment program is enacted, the utility is relieved of the need to micro-manage its customers' use. Customers are afforded the freedom and responsibility to stay in-budget as they see fit; furthermore, they don't have to comply with a detailed set of day of week or time of day rules and regulations. Those customers who want to apply more water less frequently to their trees can do to that, while others who want to use other approaches for maintaining their landscape are also free to do so. The utility can support sensible decisions by offering information on effective methods for reducing water use.

If curtailment of irrigation is required, water budgets should encourage customer investment in indoor efficiency measures as these will free up budgeted water for other uses. For example, replacement of toilets and clothes washers can generate water savings that can be used for gardening. Encouraging efficiency and flexibility is one of the strong points of water budgets.

Step 6: Identify Research Needs

During the preparation of a drought plan many areas will emerge where additional data are required. A drought plan is assembled using the best available information, but inevitable uncertainties must be identified for additional research. The drought plan should be modified as new information becomes available from the research programs.

Areas of study might include determining the baseline water use efficiency of the customers since this has a direct bearing on the potential for conservation and curtailment. It is crucial to understand the efficiency of current fixtures, appliances and process uses in order to predict available long and short range water conservation opportunities.

As water budgets are developed, areas of additional research needs quickly become apparent. For example, creating water budgets for the commercial and institutional sector is a real challenge as there is little information on what constitutes "efficient" use among these diverse end users. AwwaRF has sponsored work on the baseline demands of the most important categories of commercial and institutional water uses (Dziegielewski, et. al. 2000), but these data are not appropriate for setting water budgets as they come from a relatively small sample of customers within a limited geographical region. Furthermore, much of the data from this study came from audits conducted in the early to mid-1990s and is rapidly becoming dated. Establishing efficiency benchmarks for the CII sector remains an outstanding challenge in the field and one that is necessary for creation of effective water budgets. These data would also be useful for determining reasonable and fair water restrictions for this sector during droughts.

Step 7: Integrate Science and Policy

The goal of this step is to assure that the drought policy being formulated by elected officials is informed by the best available science on the myriad subjects involved with drought policy making. Water budgets integrate science and policy in several ways. Individualized budgets should be based on the best available science to provide each customer with allocations of water that fulfill the use intended or implied at the time that the connection fee was paid or that meet current community goals and standards. Budgets can not be set arbitrarily or they lose their validity. Water budgets can enhance drought policy formulation in the linkage between setting allocations for water use for existing customers and for the creation of a storage reserve.[18]

Step 8: Build Public Awareness

This step is critical for water budgets as well as drought planning. Public awareness is of paramount importance for the success of a drought mitigation program. Water budgets can be a powerful information tool to communicate with customers about their water use. Raising awareness on a customer level and a community level about demand in times of adequate supply makes drought communication that much easier. Public education goes hand in hand with public awareness.

Step 9: Develop Education Programs

Developing a drought education program (or a water budget education program) is an excellent way of building public awareness. A primary purpose of all such programs is to instruct the public on the important issues surrounding drought. As drought awareness is enhanced by the process of setting up water budgets, the water budget process provides an

[18] The concept of a storage reserve applies only to utilities that have the ability to store sufficient water to be used during a time of shortage. Similar to the national "Strategic Petroleum Reserve", a water provider's storage reserve is an amount of water kept in storage to be available during an emergency such as a drought or supply interruption.

excellent means for providing education on water use, water conservation, and the risk of drought.

Step 10: Evaluate and Revise Drought Plan

The final step in the drought planning process is the continual re-evaluation and revision of the drought plan. In a system with a water budget rate structure, review and evaluation of water demands are a natural part of setting water budgets for new customers and modifying budgets for existing customers. This process can keep the drought risk in public view since many utilities view drought preparation as a major objective of water budgeting. For these agencies, evaluation of the water budgets for normal use and for droughts is part and parcel with evaluation of the drought plan itself. The presence of a water budget rate structure can cause the drought plan to be revisited and updated on a regular basis.

In systems where customers pay a connection fee but never are explicitly provided with a volumetric limit for their use, the utility may be sending an unintended message: Use as much as one is willing to pay for. In water rich areas this perhaps is an intended message, but these are not the utilities that are likely candidates to implement a water budget rate structure.

Water systems with a water budget rate structure can chose to communicate that each tap comes with an expectation that only a certain volume of water is allocated to the tap. Although the developer typically pays for the tap and then has nothing to do with the water use on the site, the residents become customers of the water utility. The water budget established for the property becomes the utility's and customer's "expectation" of demand on the site. Usage above the expected amount is subject to higher rates. If water budgets are made conditional upon full supply availability for the water system, this can convey to customers the risks droughts present.

The following case study illustrates one of the earliest examples of a water budget rate structure that was implemented specifically for drought response.

DROUGHT RESPONSE CASE STUDY: EAST BAY MUNICIPAL UTILITY DISTRICT

The East Bay Municipal Utility District (EBMUD) supplies water and provides wastewater treatment for parts of Alameda and Contra Costa counties on the eastern side of San Francisco Bay in northern California. In March 1987 EBMUD began an intensive effort to curtail demand because key supply sources and reservoirs were between 50 and 60% of normal levels. EBMUD continued drought curtailment efforts through 1992 and the declared drought emergency was finally rescinded in 1993. Overall demand in the EBMUD system was curtailed between 26 – 31% through the Districts efforts (Bennett 1996).

One of the drought curtailment programs implemented by EBMUD was what the researchers believe to be the first documented example of a water budget rate structure. Created by EBMUD Water Conservation Administrator Richard "Dick" Bennett and his staff, the voluntary water budget program focused on dedicated landscape irrigation accounts such as parks, golf courses, and other large irrigated sites.

Origins

The EBMUD drought response water budget system emerged from the 1977 drought according to Dick Bennett. In 1977 landscape customers were asked to curtail use by 50% from historic average. The use of the historical average as the basis for determining compliance with the 50% cutback penalized customers that were historically efficient. EBMUD recognized this inequity in the 1977 drought response effort and when a similar 50% drought response curtailment was needed in 1988, EBMUD adopted the water budget rate structure solution.

EBMUD Rate Structure

The water budget rate structure EBMUD devised and implemented in response to the drought only impacted irrigation customers who explicitly volunteered for the program. At the conclusion of the effort in 1992, more than 1,200 landscape customers had signed on.

The water budgets were based on customer submitted measurements of landscape area. EBMUD staff randomly field verified approximately 5% of the budgets submitted and also compared the landscape area submitted to metered water use for a reasonableness test. Site surveys were conducted where the submitted data did not appear reasonable. Water budgets were constructed using 80% of the local net ET rate using a local weather station as the application depth. The landscape area and ET rate were used to calculate bi-monthly water budget volumes for each customer. The water budget was used to set the block break points in a five tier increasing block rate system and the goal was to achieve a 50% cutback in water use from the previous/baseline year.

EBMUD set the upper block rates high with the intent of penalizing customers who used more than their budget allotment. “Exceeding the target became expensive,” explained Dick Bennett. The program participants responded to the rate structure and the target cutback was achieved.

Implementation Issues

In 1988 when the program was first implemented it was not possible to modify the EBMUD billing system to accommodate the water budget rate structure. Consequently, EBMUD employed contract workers to assist with drought response to manually enter the required data into a billing spreadsheet created specifically for the program. Although implementation was labor intensive and challenging, EBMUD was satisfied with the results and kept the program in place until the end of the drought was declared in early 1993.

Future of Water Budgets at EBMUD

If drought were to revisit EBMUD in the future Dick Bennett fully expects the 1988 water budget rate structure would be dusted off and reviewed. A new $10 million customer information and billing system will come on-line in 2 years and it is anticipated that this system will be fully capable of incorporating water budgets.

Since 2002 EBMUD has offered a voluntary information landscape water budget program and definitely views this rate structure form as an option for the future.

“Landscape water budgets can provide scientifically based efficiency standards for customers,” said Dick Bennett. “You simply can’t set efficiency standards for landscapes without a water budget. In almost all cases, the landscape area needs to be determined for each meter. Efficiency standards for customers are best suited for sites with a dedicated landscape meter or where indoor use is fairly consistent as in the residential sector.”

Summary- Water Budgets and Drought Response

Water budget and water budget rate structures offer water utilities powerful new tools for reducing demand during drought. The 10 step drought planning process outlined by the National Drought Mitigation Center provides a template for combining water budgets and drought response planning that is likely to provide better overall water management than traditional drought response planning. Creation of water budgets can also be used to raise public awareness of drought risks to the community.

- A drought task force is an appropriate body to assist in the establishment of water budgets.
- Water budgets provide an excellent way to communicate the need for water conservation and proper drought response.
- The water budget rate structure lends itself to enforcement of drought restrictions in a way that does not require hit or miss drought patrols.

The water bills in a water budget rate structure inform customers of both their required use restrictions during droughts and can provide as strong an economic signal as necessary to violators to induce compliance.

CHAPTER 7

A TOOLKIT FOR IMPLEMENTING WATER BUDGET-BASED RATES

This chapter provides a toolkit for implementing water budget-based rates. The first step to implementing a water budget-based rate structure is deciding if one is needed and, if so, which kind of water budget would be most desirable. Water utilities should thoroughly assess both the potential benefits—needs that a water budget-based rate structure could potentially address—and the implementation costs. This chapter provides a set of planning tools that readers may find useful for understanding the potential benefits and the specific implementation costs.

This chapter first provides a Decision Framework (Table 7.1) for the prospective evaluation of water budget-based rate structures. If one or more needs for a water budget can be identified, the types of needs that a water budget must address will inform the process of designing a specific water budget. If water utility staff can attain a solid grasp of the potential benefits, they will be better situated to make more informed choices about budget and rate design. Thus the second step involves choosing which type of water budget is desirable to design (See Chapters 3 and 5 for guidance on budget design issues.) After a specific water budget has been conceptually designed, there are practical procedures that must be accomplished in order to *implement* a water budget-based rate structure. To address this third step, this chapter provides an implementation flow chart (Figure 7.1) to clarify the implementation process within a water utility. Understanding the mechanics of this process can help minimize implementation costs and avoid surprises.

Last, there are several predictable implementation barriers that may be encountered along the way. This chapter collects the wisdom of implementing utilities to provide lessons and insights on how these barriers can be avoided or overcome. Table 7.2 provides a tool to help identify potential implementation barriers. Early identification of potential barriers can speed the development of cost effective solutions and increase the effectiveness of water budget-based rates. The chapter concludes with a case study of a water utility that was unable to overcome all implementation barriers.

A DECISION FRAMEWORK FOR PROSPECTIVE EVALUATION

Prior to implementing a water budget-based rate structure, the first and most important implementation step is clarifying if one is needed, and if so, which one? Obtaining clarity on the desired end point can focus implementation energy where it can accomplish the most. Table 7.1 provides a Decision Framework to structure this decision process through an identification of needs. By carefully working through the questions attached to each type of need, users can better define exactly which kind of water budget would address the needs of the local conditions. The Decision Framework identifies the following potential needs for a water budget:

- Need to Address Long Term Supply Limitations
- Need for Improved Drought Management
- Need for a More Effective Rate Structure
- Need for Better Water Management
- Need for Improved Customer Service

There are questions provided for each of these types of needs that, in turn, point to possible applications of water budgets. After carefully working through the Decision Framework, users should have some idea of the potential benefits of applying a water budget or a water budget based rate structure. If the potential benefits can be identified, there is then cause to proceed through the flow chart in Figure 7.1.

IMPLEMENTATION PROCESS FOR WATER BUDGET-BASED RATE STRUCTURES

The implementation flow chart in Figure 7.1 describes the implementation steps required once an initial decision (diamond) has been made to pursue a water budget application. The flow chart describes the conceptual definition of the water budget application (either a water budget stand-alone program or a water budget-based rate structure) as the next implementation step. This conceptual design should then be reviewed and approved by internal management—strong consensus at this point can be critical. There follow a series of line staff functions:

Line staff functions

- Define water budget
  - Define data and information system requirements
  - Assess alternative budget definitions
- Design and test billing system
  - Redesign customer billing statement
- Public involvement
  - Initial customer communication
- Customer outreach
  - Assess ongoing customer service
  - Augment customer outreach

The final approval step is that of obtaining external approval (typically from an elected board or council) if a water budget based rate structure is involved.

Table 7.1 Decision framework for prospective evaluation of water budget rate structures

Needs Identification	Questions to Answer	Potential Application
Need to Address Long Term Supply Limitations	• Is your water utility currently confronting long term water supply limitations? • Is your water utility (even if in a situation of water sufficiency) facing risks and uncertainties to water supplies?	Water budgets as a means of improving the targeting of other/existing conservation programs. Water budget-based rate structures (WBBRs)
Need for Improved Drought Management	• Has your water utility implemented customer demand curtailments in the recent past? • Has your water utility faced customer acceptance issues when implementing curtailments? • Has revenue management been an issue during droughts?	Water budgets in standby mode WBBRs can allocate water shortages in ways customers find more acceptable WBBRs can improve revenue management during curtailment events
Need for a More Effective Rate Structure	• How effective is your utility's current rate structure in terms of revenue generation, customer service, demand management, drought response?	Water budgets as a means of helping agencies and customers better manage their water use.
Need for Better Water Management	• Do water costs exceed benefits (economic case for water use efficiency)? • Are there issues providing water service to outdoor irrigation uses? • Are there operational issues with system peaks?	Water budgets as a stand alone conservation program. Water budgets as a means of improving the targeting of other/existing conservation programs. Water budgets as a peak reduction program.
Need for Improved Customer Service	• How happy are your customers with your utility's: ○ responsiveness to bill complaints? ○ information on using water wisely? ○ information on efficient landscape use?	Water budget programs can provide more tangible and understandable information on efficient and appropriate water use. Water budgets offer direct feedback on what constitutes efficient use.
Note: These potential needs are not mutually exclusive. Water budget applications identified from this table still need to have their benefits and costs assessed. See Implementation process flow chart – Figure 7.1.		

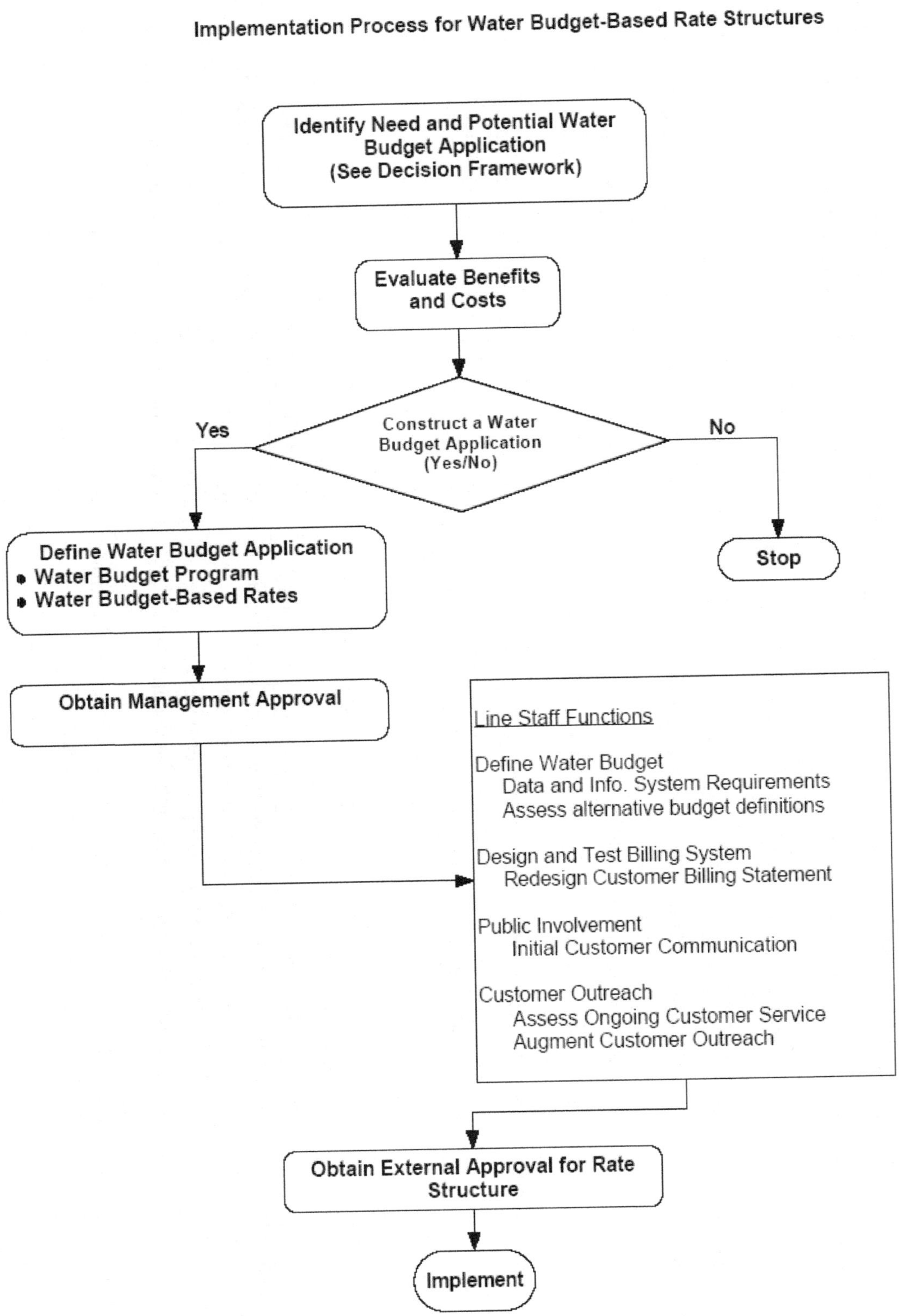

Figure 7.1 Water budget-based rate structure implementation flow chart

BARRIERS TO IMPLEMENTATION OF WATER BUDGET-BASED RATES

Water budget-based rates are probably not for everyone. Although the basic idea for these rate structures has been around for nearly twenty years and interest has been steady, until recently only a handful of agencies had ever successfully implemented a water budget-based rate structure. Even if a water utility is interested in implementing this type of rate structure, a number of real and perceived barriers may delay or block implementation from moving forward. The extent to which these barriers pose real or imagined impediments to implementation depends greatly upon the utility, the staff, the public, the water supply situation, and the local decision makers. The examples of successful implementation of water budgets described in this report show that implementation is possible and that all of these barriers can be overcome with the right combination of know how, exigent circumstances, and political will[19]. Understanding potential barriers to implementation can be helpful when developing a strategy for moving forward with a new rate structure concept.

Some of the perceived barriers to adopting and implementing a water budget-based rate structure include:

- Complexity
- Data requirements
- Software requirements
- Lack of local precedent
- Equity concerns
- Cost of service concerns
- Revenue requirements
- Institutional resistance
- Political resistance

In some cases these barriers can be overcome, in others they cannot. A brief discussion of each of the barriers noted above follows in Table 7.2. An example of a failed attempt to implement water budgets in West Jordan, Utah is presented as a case study. The extent to which each of these barriers may present a problem depends entirely on the situation present at each local agency.

[19] The authors were not able to identify any investor-owned utilities that have implemented water budgets. It seems likely that the regulatory environment of investor-owned utilities will pose additional barriers to implementing water budget rates.

Table 7.2 Barriers to implementing a water budget-based rate structure

Potential Barrier	Questions to answer	Implication
Data requirements for developing defensible water budgets	• Do your conservation programs use water budgets for any outdoor water use efficiency programs? • Do you have or can you obtain access to public information (tax assessor data) on lot size, home size, etc.? • Can parcel-level data be obtained from your GIS?	• Data required for water budgets may already exist in local databases or GIS. • Consistent customer-specific data may be available at low cost.
Customer billing system requirements	• Can the current customer billing system implement an increasing block rate structure? • Is the current customer billing system using software capable of relational database functions?	• Current billing system may have sufficient capability. • Many customer billing systems do have functional capability. • Typically only a few calculated fields are required for water budget-based rate structure implementation.
Compatibility of water budgets with cost-of-service principles	• Does your utility have customer classes defined? • Does the current rate structure differentiate by customer class? • Does the current rate structure implicitly or explicitly allocate revenue requirements by customer class?	• Customer class definitions are required for customer class-based rates. • Cost-of-Service practices do not require cost allocation by individual customer, only customer class
Pre-implementation customer communication	• Are any of your customers familiar with the concept of water budgets? • What kind of customer involvement process has been used in the past for rate reform? • Are citizen representatives part of the rate structure review process? • Is the press covering the rate structure issue?	• It is important to understand beginning level of customer understanding and community values • Customers may already be familiar with the concept of water budgets. • Public involvement in rates is increasingly important (See AWWA 2004 *Avoiding Rate Shock*)
Ongoing customer communication	• Does your customer bill currently communicate information on appropriate levels of water use? • What are your current staffing levels for answering and responding to customer queries?	• Customer bill redesign is often required • This is both a customer service issue (responding to bill complaints) and a conservation outreach issue.

Note: This table is predicated on the assumption that a need/desire for water budget-based rates has been established and that the prospective benefits can be greater than prospective costs. Early identification of potential barriers can speed the development of cost effective solutions and increase the effectiveness of water budget-based rates.

Barrier Case Study: West Jordan, Utah

One of the barriers to implementing any major change in city government is the lack of local precedents to demonstrate the success of that change. In West Jordan, Utah water budgets were seriously considered but ultimately not adopted when a number of the barriers discussed above were encountered. West Jordan has a water service population of approximately 86,000 people and significant growth potential. West Jordan receives most of its water from the Jordan Valley Water Conservancy District. It delivers about 17,000 acre-feet of water per year to a customer base that includes residential, commercial, and two large industrial customers. West Jordan currently uses an increasing block rate billing structure.

Budget based rate structures were introduced to West Jordan in 1999 at the recommendation of the city's Water Conservation Committee. Dubbed "Target Billing" by the city's Water Conservation Coordinator, the water budget concept enjoyed some acceptance at first but eventually became entangled in scientific terminology and formulae, privacy concerns, inaccurate meter readings, focus groups, and city politics.

In an attempt to gain support for the water budget concept, Tom Ash a consultant who had participated in implementing a similar rate structure at the Irvine Ranch Water District in California, was invited to make a presentation to the West Jordan City Council. After his appearance several West Jordan residents who had considerable experience in water issues, including the city's Water Conservation Committee, voiced support for a change to a water budget rate structure format suggesting that it would place the city at the forefront of the growing conservation movement in Utah. The City Council directed staff to bring back an ordinance to implement target billing in three weeks. The motion passed five to one, with one council member absent.

At a later meeting of the city council, and after staff consultation with another water district that had recently adopted a budget based rate structure, a simplified version was presented. Again, a motion was made to direct staff to complete the final research and development of the simplified water budget program as outlined by staff, and to bring back an ordinance for implementation. That motion passed six to zero with one absent.

At this point the water budget based rate structure appeared on its way to implementation in West Jordan but several issues were dogging the process. The first barrier to acceptance by the city council was understanding and accepting the scientific terminology and formulas associated with water budgets. Evapotranspiration was a concept clearly understood by members of the committee that recommended target billing be adopted, and perhaps they underestimated the degree of difficulty that council members and billing staff would have in grasping this essential concept. Formulae for defining the budget for each customer caused additional consternation.

Another barrier encountered in West Jordan was the choice to request residential customers to disclose the number of people in the home as the basis for the indoor component of the water budget. This raised numerous privacy concerns among city council, staff and some residents. Among those concerns were:

- The possibility of crafty people "gaming the system" and reporting more people than were actually living at the residence

- The perception that normal family dynamics would create overwhelming customer service problems, such as empty nesters that have family members moving back home or visiting for an extended period
- Submitting a request to the city for an increase in the indoor budget every time a new baby was born.

During this incubation period for water budget based rates the billing department experienced an inordinate number of erroneous reads of the water meters in West Jordan. About 5 percent of customer bills required adjustment. With the implementation of budget based rates, inaccurate reads could throw the unlucky customers into the penalty rate creating a customer service problem. After these problems cropped up, the meter reading function was brought in-house and new radio read meters were installed throughout the system.

To address the issues and concerns raised about water budgets, West Jordan decided to hire a consultant to conduct a focus group process to determine how city customers would react to implementation of the simplified water budget based rate structure. The new rate structure was presented as a reformatted water bill showing the base charge, graduated commodity charges, the amount of water allocated for the budget, and penalty blocks based on the percentage of water used beyond the budgeted amount. Participants liked the new billing format but raised questions about the advantages of budget based, as opposed to increasing block rates. Among the issues raised was the possibility that small lot owners would be the primary payers of rate penalties brought about by the shift from increasing block rates to water budget based rates.

Another issue raised in the focus group concerned the accuracy and sustainability of the data used to develop the water budgets. For example, staff applied sampled average irrigated area values (similar to what was done by Centennial Water and Sanitation District), but the focus group expressed concern that the averaged areas might not accurately reflect differences between lots in a specific way. Two identical lots side by side could differ significantly in the amount of irrigated landscaping present if one had a paved RV pad and the other did not. Yet under the West Jordan rate plan, these customers would be given the same water allocation. West Jordan could not identify practical ways to resolve these issues.[20] Several members of the focus group felt that simplicity was an important attribute of a water rate system and that the proposed water budget rates were anything but simple.

The net result of the process was that West Jordan could not overcome the barriers identified and opted not to implement a water budget-based rate structure.

---

[20] Minutes of city council meeting dated April 27, 2004

CHAPTER 8

CONCLUSIONS AND RECOMMENDATIONS

"Water budgets"—volumetric allotments of water to customers based on customer-specific characteristics and conservative resource standards—are an innovative means of improving water use efficiency. Once thought to be impractical because of technological constraints, water budgets linked with an increasing block rate structure have been implemented successfully in more than 20 utilities. As utilities develop advanced customer information systems and geographical information systems these rate structures are expected to be applied more broadly. Water budget rate structures are attractive to water agencies searching for stable revenue generation, improved customer acceptance, increased water use efficiency, augmented affordability of nondiscretionary customer uses, and improved drought response. A growing number of agencies believe that water budget rate structures offer potential benefits to water utilities and their customers in coping with increasing water scarcity.

Water budget-based rate structures are probably not for everyone. This rate structure form requires more customer level data than traditional rate structures and depending upon the implementation circumstances may be more expensive and labor intensive to put in place. Agencies with little or no interest in water conservation are not good candidates for water budgets since this rate structure form has been shown to be an effective conservation tool. For utilities that are serious about water conservation and sending a fair and effective price signal about usage to their customers, water budgets appear to be an excellent option to consider.

This research project examined water budgets and their potential value to North American water utilities and the varying applications of the water budget concept that have been adapted to different conditions. Key issues identified include: different practical approaches to water budget rate structures, the benefits and challenges of these approaches, the potential uses of water budgets during drought, and important steps in the water budget implementation process. The case studies presented in this report offer first hand information on the design, implementation, and management of water budget rate structures as well as some cautionary examples of water budget programs that were rescinded or that never got off the ground.

RESEARCH CONCLUSIONS

Although water budgets have been implemented by water providers in North America for nearly 20 years, until recently they had only been implemented by a handful of utilities. As populations increase and climate uncertainties place heightened demand and stresses on water systems, new tools for water conservation and drought response are sought. Water budget-based rate structures are one of the most effective tools available to water providers to provide a meaningful price signal, reduce water waste, increase efficiency, and manage drought response in a sensible and equitable manner. A review of the research conducted for this study resulted in a number of conclusions regarding water budget-based rate structures.

Definition of a water budget

A water budget is a formal definition of the quantity of water that would be required by an efficient level of water use. Exactly what constitutes an "efficient level" of use (particularly as it relates to outdoor irrigation) is often a community decision and must be based upon public expectations as well as water supply realities. Water providers tend to define water budgets in more specific terms so that their customers can easily understand how the budget (or allocation as it is often called) applies. For example, the City of Boulder, CO defines the concept for customers in the following way: "A water budget is the amount of water you are expected to need for a specific month. Each customer's water budget will be different based on their water needs. Water budgets may vary monthly based on seasonal outdoor watering needs."

Landscape water budgets are the most common form of a water budget. A landscape water budget is typically a volume of water that is calculated from two fundamental parameters: the landscape size (usually in square feet) and the water requirement of plants in that area which is typically represented by the evapotranspiration (ET) rate.

Indoor water budgets are typically a fixed volume provided each month based on the historic use of the customer class or a calculated volume based on the anticipated needs of the customer.

Water Budget-Based Rates

Water budget-based water rates—also known as individualized, goal-based, and customer specific rates—are block rates where the block is defined by using one or more customer characteristics. Water budget-based rate structures can be thought of as an increasing block rate structure where the block definition derives from a water budget that defines an efficient level of water use for each customer.

Water Budgets Can Meet Cost of Service Requirements

According to the M1 Manual, "the basic premise in establishing adequate rate schedules that are equitable to different customers is that rates should reflect the cost of providing water service" (AWWA 2000). After revenue requirements have been established, costs are allocated among different types of water users, and then rates are designed to reflect the cost of providing water service. Although the concept of cost of service is central to public utility rate setting, most rates reflect historical pricing regimes and local political imperatives. Legal precedence requires that rates be "fair and reasonable." Public utility practice often uses a cost-of-service justification to help determine.

A water budget-based rate structure is inherently no more or no less cost-based than a similar increasing block rate structure without a water budget component. There is little dispute that increasing block rate structures can be properly designed to meet cost-of-service requirements. The same is true for a water budget-based rate structure. The research team could find no report or publication that identified cases where water budget rate structures failed to meet cost of service criteria. Given that more than 20 utilities are actively implementing some form of these rate structures, it appears that properly designed water budget-based rate structures can meet these important criteria.

Water Budgets are Being Used Across the USA

During the 1990s, water budget rate structures were only implemented by a few water agencies in California, most notably in Capistrano Valley, IRWD, and the Otay Water District. By 2007, water budget rate structures have diversified and expanded throughout California and across the United States to Utah, Nevada, Colorado, North Carolina, and Florida. The researchers were able to identify 25 water providers that are using some form of a water budget for billing and/or informational purposes.

While still primarily a California and western states phenomena, water budgets have found support in a broad range of agencies from moderate-sized systems like Marco Island, Florida to very large utilities like Los Angeles Department of Water and Power (LADWP). Water budgets are no longer an exclusive "boutique" rate structure for select technologically savvy providers. These rate structures have been adopted by a broad range of providers, each adopting their own version of the water budget concept.

Water Budget-Based Rate Structures, Like Utilities, are Unique

Utilities are unique and so are their rate structures be they water budget-based or not. The formulation of water budgets and implementation within the utility context was unique for every example the researchers identified. Only three of the identified water budget utilities developed a water budget for all customers in the service area. Most of the water budget utilities opted to use water budgets for only certain customer classes (such as single-family residential or dedicated irrigation) and some only use water budgets for informational purposes in which the budget is not linked to the rate structure. Single-family residential and dedicated irrigation accounts were the most common customer categories to be included in a water budget program.

Landscape Water Budgets Are Most Frequently Implemented

Landscape water budgets were the original basis for the formulation of these rate structures and are the foundation of the majority of implementations examined in this report. The idea of calculating the volume of water required to adequately irrigate a plot of land is hardly a new concept. Farmers and agronomists have been developing water budgets for agriculture for many years using evapotranspiration measurements to determine the required application depth to maximize plant growth. In the urban context, water budgets have been used for years to determine irrigation efficiency for parks and large irrigated parcels. They have been used at least since 1989 in conjunction with inclining block rate structures to provide a tailored price signal based on the water requirements of the landscape.

Water Budget-Base Rate Structures Require a Capable Utility Billing System and Customer Level Data

The implementation of water budget-based rate structures has been made possible by the advent of computerized utility billing systems that can incorporate specific customer-level information into a billing calculation. None of the water budget formulas or calculations studied by the research team was particularly complex or complicated. Most modern, database centered, utility billing systems can probably be adapted to incorporate water budgets without significant effort. Billing systems based on older technology may not be capable of handling water budgets.

At least one utility billing software company has developed a module specifically for incorporating water budget-based billing.

The development of most water budgets examined in this study requires some customer level data such as lot size or irrigable area. Many water utilities have invested in geographical information systems that provide parcel (or customer) level data such as lot size and pervious/impervious area measurements. If these parcel level data can be linked to the customer billing system, then much of the data required to implement a landscape water budget program are already in place. Utilities that do not have a GIS may be able to use local tax assessor records to obtain lot size information and other data that could be used to develop water budgets.

Implementation Costs Vary

The costs associated with implementing a water budget-based rate structure vary tremendously depending upon the circumstances of the water agency involved. Most water budget systems have been implemented "in-house" using utility staff and limited outside expertise. Often the existing customer billing system can be modified to accommodate the new rate structure. In these situations, the implementation costs amount only to the staff time required to develop the program. IRWD, San Juan Capistrano, Otay, and Centennial all implemented their water budget programs in under 12 months with most labor coming almost entirely from existing staff resources.

Some water budget implementations are more expensive. The City of Boulder spent nearly three years studying and then implementing its water budget rate structure. Boulder first spent several hundred thousand dollars on a study to investigate the feasibility of water budgets. Boulder's antiquated billing system (hardware and software) was replaced as part of the water budget implementation at a cost of more than $1,000,000. It should be noted that Boulder intended to replace this system anyway and had budgeted money for the project prior to the decision to implement water budgets. Boulder also commissioned a cost of service study as part of the rate implementation which increased the overall cost of the project.

Water Budget-Based Rate Structures Spur Conservation Savings

Many of the utilities that have implemented water budget-based rate structures have experienced substantial conservation savings attributable to the rate structure and accompanying customer outreach programs. Analysis by Irvine Ranch Water District (IRWD) staff concluded that the program increased outdoor efficiency by 60 percent. In 1992 shortly after the rate structure was implemented IRWD customers irrigated approximately 3,600 acres in the service area and applied roughly 4.4 acre-feet of water per acre (52.8 inches of water applied). The typical ET requirement for turf grass in the area is 48 inches per year. In 2005 IRWD customers irrigated 11,768 acres of landscape and applied only 1.7 acre-feet of water per acre (20.4 inches). This represents a 61% reduction in the outdoor application rate since the introduction of the water budget program. An independent evaluation in 1997 documented a 37% decline in water consumption as a result of IRWD's rate structure and customer outreach (Pekelney and Chesnutt, 1997).

The same 1997 study found that two other water budget rate structures and conservation programs in California also achieved substantial conservation savings. San Juan Capistrano's water budget rate structure and customer outreach reduced raw consumption by 35% between the pre- and post-program periods; while the Otay Water District experienced a 20% decline in usage, after controlling for weather (Pekelney and Chesnutt, 1997).

Staff from the Centennial Water and Sanitation District in Colorado documented an average 25% reduction in demand after implementing a water budget program. Even though Centennial continues to add customers at a fairly steady rate, its demand has decreased 18 – 31% starting in 2003 when the water budget-based rate structure was implemented.

Most Water Budgets Are Seen As "Fair"

One of the key benefits of water budgets identified by the implementing water agencies studied for this project was their greater perceived "fairness". Most of the agency staff involved said the additional complexity of customer-specific water budgets was more than outweighed by the increased customer acceptance of the customized rate structure. Staff found that once customers understood the system customers preferred to have their rates based on the characteristics of their site rather than on an arbitrary or average value. Variance programs have also been used to increase the accuracy of both indoor and outdoor water budgets since the water agency cannot know about every circumstance of each individual customer.

Landscape Penalty Rate Structures Are Less Successful

Landscape water budget systems based on a penalty rate structure, such as the one implemented by Otay in 1992 and rescinded in 2000, are more problematic than the water budget-based rate structures implemented in other cities. The Otay system relied on self reported landscape information and issued first a warning and then penalties for exceeding budgeted allotments. Although this approach resulted in a 20% reduction in demand among the budgeted customers, the rate penalty system proved problematic with some customers. The system generated numerous customer complaints and the use of self-reported landscape area data allowed some customers to inflate their area estimates and defeat the intent of the system.

To be successful, water budgets must be based (to the extent possible) on measured customer level data, not self-reported information. Water budgets are also more likely to gain acceptance if they are connected with a well designed and communicated rate structure rather than a penalty structure.

Water Budgets Offer Great Potential as a Drought Response Tool

Water budgets and water budget rate structures offer water utilities powerful new tools for reducing demand during drought and for monitoring customer compliance with drought restrictions. In time of shortages, water budgets allow a water provider to quickly and easily identify excess use and even penalize it if necessary. By summing all water budgets, utilities can quickly understand the amount of water likely to be required to meet customer demands in any given billing cycle. During a drought, water budgets have the potential to assist water utilities in more fairly apportioning demand reductions among customers with different needs and among different customer classes since the reference point for reductions is based on the water required by each customer in normal times.

Water budget rate structures can help with drought plan enforcement in the area of communications. The water budget rate structure, with its billing system, informs all customers on a regular basis of the required use reductions. The water bill can show each customer how much water she is allocated during the drought. This information can theoretically be developed well before the drought occurs as part of the budgeting process. This is a far more reliable and effective way to implement drought related conservation since it is pre-planned rather than improvised. The billing system is already in place and the bills can provide the public with the information needed to respond to the drought.

Another way that water budget rate structures aid with drought plan implementation is in the enforcement of mandatory demand curtailment. A simple query of the billing database can inform the utility each billing period which customers have complied with drought restrictions and remained within budget and which have not. If the higher water rates being charged are not sufficient to elicit co-operation then additional fines and penalties can be considered. This is a highly reliable system. Unlike the "water cop" approach where customers are ticketed if they happen to be observed violating the drought restrictions, a water budget drought enforcement program automatically identifies every customer who is not complying, thus enabling fair and uniform enforcement. Water enforcement patrols are costly and can only catch violators "in the act" of violating a watering restriction. A water budget, however, provides a regular and automatic check on which customers are in or out of compliance with drought response.

RECOMMENDATIONS

The recommendations from this research project are divided into two categories: (1) recommendations for water agencies considering water budgets; and (2) recommendations for future research.

Recommendations for Water Utilities Considering Water Budgets

Decide if a Water Budget Rate Structure Makes Sense for the Organization

Water budget-based rate structures are not for everyone. This rate structure form requires more customer level data than traditional rate structures and depending upon the implementation circumstances may be more expensive and labor intensive to put in place. Agencies with little or no interest in water conservation are not good candidates for water budgets since this rate structure form has been shown to be an effective conservation tool. For utilities that are serious about water conservation and sending a fair and effective price signal about usage to their customers, water budgets appear to be an excellent option to consider.

Commitment, Coordination and Education are Essential

Designing and implementing a water budget-based rate structure involves all levels of a water agency including the governing board, general manager, rates staff, conservation staff, accounting and finance staff, computer and information systems, public affairs, and the customers. To maximize the potential for success, coordination, communication, and education with all of those involved in the project is required. A thorough understanding of the motivation and operation of the water budget system by all levels of the organization will increase effectiveness and allow the program to be improved over time.

The board and staff must be firmly convinced the water budget rate program is a good idea if they are to stand behind it. The concept should be well developed and adjusted until it is well understood and supported. Changes in rate structure almost always generate grumbling from some segments of the customer base. With sustained commitment to the water budget program, leaders of the agency can successfully support and defend the effort and will be prepared and motivated to fix problems when they are identified.

Identify Information System Requirements Early

The implementation of water budget-based rate structures requires computerized utility billing systems that can incorporate specific customer-level information into a billing calculation. Most modern, database centered, utility billing systems can probably be adapted to incorporate water budgets without significant effort, but it is important to identify data and billing system requirements early on. This will enable the utility staff to develop or purchase any necessary hardware and software and to obtain the data necessary to implement the program in time for the launch date. A number of agencies studied in this project stated that the critical ingredient for success with a water budget system was not hardware or software; it's "peopleware" – i.e. the people who make the program happen.

Educate and Inform

When implementing a water budget rate structure, educating and informing customers about the rate structure and how it works is essential. Customers need to understand not only how the rate structure works and will impact them, but also why it is being implemented. The City of Boulder sent out at least four mailings to customers in advance of implementing the new water budget rate structure. In addition there were articles in local newspapers and extensive information on the City's web site.

Respond to Customer Concerns

No water budget program is perfect. The water utility and its board should be ready and willing to make adjustments to the program based on good suggestions from customers. Many suggestions will likely come when the program is first implemented, but others will come through the fullness of time and experience. Adaptation through years of implementation has been a hallmark of the successful water budget programs in San Juan Capistrano and IRWD. Budget adjustments and variances with reasonable cause are important for achieving customer acceptance and buy-in to the program. In a customized rate structure, customers expect to be treated as individuals. The water utility should be ready with staff and information to respond to customers' questions and concerns right from the outset of the program.

Use Scientific Basis for Water Budgets Whenever Possible

Water budgets should be objective and must make sense to customers. To the extent possible they should be based on empirical data (indoor budgets) and horticultural science (outdoor budgets). Be explicit about the formulation of the allocations and be up front about areas where agency discretion enters into the allocation formula.

Time Program Implementation Strategically

New rate structures are usually best implemented during off-peak season. This allows customers to adjust to the new rate structure for several billing periods before encountering the potential "sticker shock" in the peak season.

Take Advantage of the Information Provided by the Water Budget System

Water budgets have tremendous potential as a tool to assist with supply shortages such as drought. They also provide an easy means for targeting water conservation programs to the customers who can most benefit from them. If budgets are set properly, then customers who grossly exceed their budget become prime candidates for targeted conservation measures. Standard increasing block rate structures don't provide this level of insight into customer demands.

Drought response may prove to be the most important benefit of water budgets. In order to make the most of the potential benefits, utilities must understand how to take advantage of the rate structure to encourage and, if necessary, mandate demand curtailment. During a drought the utility can adjust the pricing tiers and the budget size as necessary to achieve the desired level of curtailment.

Target Additional Incremental Revenue to Incremental Water Efficiency Programs

IRWD uses all of the revenue from all tiers above budget allocation to fund additional water conservation measures, water reuse, and environmental programs. This has proved to be a strong selling point for its water budget program. Additionally IRWD's rates are among the lowest in Orange County, CA. Customers like the idea of the utility investing the high tier payments in cost-justified water efficiency programs. This revenue can go towards subsidizing the low tier rate, providing rebates for water efficient fixtures, hardware, and equipment, and for conservation staff. This in turn results in reduced water use and lower bills for customers. Because IRWD covers fixed costs on a separate part of the bill, reduced demand does not negatively impact utility revenue or operations. This linkage helps explain where the money for water conservation programs comes from and how the rate structure contributes to conservation beyond just sending an effective price signal.

Recommendations for Future Research

Advanced Tools for Water Budget Calculation and Implementation

Water agencies wishing to implement a water budget-based rate structure are essentially on their own when it comes to developing a customer billing system capable of handling the rate structure and when it comes to developing water budgets for diverse classes of customers. Future research could tackle these issues by specifying billing system requirements for implementing water budgets. This task was accomplished to a small degree in this project, but could be greatly expanded.

Establishing accurate and reasonable water budgets remains a challenge, particularly for the CII sector. Future research could explore methods for using GIS and other geospatial technologies for measuring irrigated area remotely. This could help automate the process for establishing landscape water budgets.

For the CII sector, additional benchmarking data are required to determine efficient usage levels from different classes of customers. Current CII budgeting methods that use historic consumption to set future budgets are ineffective at developing budgets that encourage efficiency. Benchmarking measures based on key usage parameters such as number of employees, meals served, cars washed per day, building area, etc. could be a more effective method for establishing water budgets. To accomplish this, additional data are required. Such an effort could be combined with an update of the AwwaRF *Commercial and Institutional End Uses of Water* Study, originally published in 2000.

Water Budgets and Drought Response

This project has identified drought response as an area where water budget-based rate structures offer tremendous potential benefit. However, utilities have only limited field experience with managing water shortages with the aid of water budgets. In the coming years this is very likely to change. An important area of future research will be to examine and evaluate the uses of water budgets during real water shortage situations.

New Implementations of Water Budget-Based Rate Structures

Although the water budget concept is not new, over the past five years the number of agencies implementing water budgets has grown from five to twenty-five. In the coming years it is likely that more water providers will move towards this type of conservation oriented rate structure. This research study represents the beginning of objective analysis into the implementation and performance of water budget-based rate structures. As new ideas and concepts are put into place, this research should be updated and expanded in the hope of improving understanding of water budget-based rates and how they can be applied in the water utility context.

APPENDIX A

CALIFORNIA AB 1881

Assembly Bill No. 1881
CHAPTER 559
An act to add Section 1353.8 to the Civil Code, to repeal and add Article 10.8 (commencing with Section 65591) of Chapter 3 of Division 1 of Title 7 of the Government Code, to add Section 25401.9 to the Public Resources Code, and to add Article 4.5 (commencing with Section 535) to Chapter 8 of Division 1 of the Water Code, relating to water conservation.
[Approved by Governor September 28, 2006. Filed with Secretary of State September 28, 2006.]

legislative counsel's digest
AB 1881, Laird. Water conservation.
(1) Existing law, the Davis-Sterling Common Interest Development Act, defines and regulates common interest developments, which include community apartment projects, condominium projects, planned developments, and stock cooperatives. This bill would provide that the architectural guidelines of a common interest development shall not prohibit or include conditions that have the effect of prohibiting the use of low water-using plants as a group.

(2) The Water Conservation in Landscaping Act requires the Department of Water Resources to appoint an advisory task force to work with the department to draft a model local water efficient landscape ordinance that local agencies may adopt, requires the task force to submit the ordinance to the department on or before May 1, 1991, and requires the task force to cease to exist on the date the department adopts the model ordinance or January 1, 1992, whichever occurs first. The act requires the department, not later than January 1, 1992, to adopt a model local water efficient landscape ordinance which each local agency may adopt. The act makes the model local water efficient landscape ordinance adopted by the department applicable within the jurisdiction of a local agency if that local agency, by January 1, 1993, has not adopted a water efficient landscape ordinance or has not adopted certain findings that the adoption of the ordinance is unnecessary. This bill would specify that the provision making the model ordinance applicable to a local agency on and after January 1, 1993, does not apply to chartered cities. The bill would require the department, to the extent funds are appropriated, not later than January 1, 2009, by regulation, to update the model ordinance in accordance with specified requirements. The bill would require the department to prepare and submit to the Legislature a prescribed report before the adoption of the updated model ordinance. The bill would require a local agency, not later than January 1, 2010, to adopt the updated model ordinance or other water efficient landscape ordinance that is at least as effective in conserving water as the updated model ordinance. The bill would make the updated model ordinance applicable within the jurisdiction of a local agency, including a chartered city, if, by January 1, 2010, the local agency has not adopted its own water efficient landscape ordinance or the updated model ordinance. The bill would require each local agency, not later than January 31, 2010, to notify the department as to whether the local agency is subject to the department's updated model ordinance and, if not, to submit to the department a copy of the water efficient landscape ordinance adopted by the local agency, among other documents. The bill would require the department, to the extent funds are appropriated, not later than January 31, 2011, to prepare and submit a report to the Legislature relating to the status of water efficient landscape ordinances adopted by local agencies. By imposing requirements on local agencies in connection with the adoption of water efficient landscape ordinances, the bill would impose a state-mandated local program.

(3) Existing law requires the State Energy Resources Conservation and Development Commission (Energy Commission), after one or more public hearings, to take specified action to reduce the wasteful, uneconomic, inefficient, or unnecessary consumption of energy. Existing law requires the Energy Commission, by January 1, 2004, to amend specified regulations to require that residential clothes washers manufactured on or after January 1, 2007, be at least as water efficient as commercial clothes washers, and to take certain other related action. This bill would require the Energy Commission, in consultation with the department, to adopt, to the

extent funds are available, by regulation performance standards and labeling requirements for landscape irrigation equipment, including irrigation controllers, moisture sensors, emission devices, and valves to reduce the wasteful, uneconomic, inefficient, or unnecessary consumption of energy or water. The bill would require the Energy Commission to adopt those requirements for landscape irrigation controllers and moisture sensors by January 1, 2010, and, on and after January 1, 2012, would prohibit the sale or installation of an irrigation controller or moisture sensor for landscape use unless the controller or sensor meets those adopted requirements. The bill would require the Energy Commission, on or before January 1, 2010, to prepare and submit to the Legislature a report that sets forth a proposed schedule for adopting performance standards and labeling requirements for emission devices and valves.

(4) Existing law generally requires an urban water supplier to install water meters on all municipal and industrial service connections located within its service area on or before January 1, 2025. This bill would require a water purveyor as defined, to require as a condition of new retail water service on and after January 1, 2008, the installation of separate water meters to measure the volume of water used exclusively for landscape purposes. The bill would make this requirement applicable to specified service connections.

(5) The California Constitution requires the state to reimburse local agencies and school districts for certain costs mandated by the state. Statutory provisions establish procedures for making that reimbursement. This bill would provide that, if the Commission on State Mandates determines that the bill contains costs mandated by the state, reimbursement for those costs shall be made pursuant to these statutory provisions.

The people of the State of California do enact as follows:
SECTION 1. Section 1353.8 is added to the Civil Code, to read:

1353.8. The architectural guidelines of a common interest development shall not prohibit or include conditions that have the effect of prohibiting the use of low water-using plants as a group.

SEC. 2. Article 10.8 (commencing with Section 65591) of Chapter 3 of Division 1 of Title 7 of the Government Code is repealed.

SEC. 3. Article 10.8 (commencing with Section 65591) is added to Chapter 3 of Division 1 of Title 7 of the Government Code, to read: Article 10.8. Water Conservation in Landscaping 65591. This article shall be known and may be cited as the Water Conservation in Landscaping Act.

65592. Unless the context requires otherwise, the following definitions govern the construction of this article:
(a) "Department" means the Department of Water Resources.
(b) "Local agency" means any city, county, or city and county, including a charter city or charter county.
(c) "Water efficient landscape ordinance" means an ordinance or resolution adopted by a local agency, or prepared by the department, to address the efficient use of water in landscaping.

65593. The Legislature finds and declares all of the following:
(a) The waters of the state are of limited supply and are subject to ever increasing demands.
(b) The continuation of California's economic prosperity is dependent on adequate supplies of water being available for future uses.
(c) It is the policy of the state to promote the conservation and efficient use of water and to prevent the waste of this valuable resource.
(d) Landscapes are essential to the quality of life in California by providing areas for active and passive recreation and as an enhancement to the environment by cleaning air and water, preventing erosion, offering fire protection, and replacing ecosystems lost to development.
(e) Landscape design, installation, maintenance, and management can and should be water efficient.

(f) Section 2 of Article X of the California Constitution specifies that the right to use water is limited to the amount reasonably required for the beneficial use to be served and the right does not and shall not extend to waste or unreasonable use or unreasonable method of use.
(g) (1) The Legislature, pursuant to Chapter 682 of the Statutes of 2004, requested the California Urban Water Conservation Council to convene a stakeholders work group to develop recommendations for improving the efficiency of water use in urban irrigated landscapes.
(2) The work group report includes a recommendation to update the model water efficient landscape ordinance adopted by the department pursuant to Chapter 1145 of the Statutes of 1990.
(3) It is the intent of the Legislature that the department promote the use of this updated model ordinance.
(h) Notwithstanding Article 13 (commencing with Section 65700), this article addresses a matter that is of statewide concern and is not a municipal affair as that term is used in Section 5 of Article XI of the California Constitution. Accordingly, it is the intent of the Legislature that this article, except as provided in Section 65594, apply to all cities and counties, including charter cities and charter counties.

65594.

(a) Except as provided in Section 65595, if by January 1, 1993, a local agency did not adopt a water efficient landscape ordinance and did not adopt findings based on climatic, geological, or topographical conditions, or water availability that state that a water efficient landscape ordinance is unnecessary, the model water efficient landscape ordinance adopted by the department pursuant to Chapter 1145 of the Statutes of 1990 shall apply within the jurisdiction of the local agency as of that date, shall be enforced by the local agency, and shall have the same force and effect as if adopted by the local agency.
(b) Notwithstanding subdivision (b) of Section 65592, subdivision (a) does not apply to chartered cities.
(c) This section shall apply only until the department updates the model ordinance.

65595.

(a) (1) To the extent funds are appropriated, not later than January 1, 2009, by regulation, the department shall update the model water efficient landscape ordinance adopted pursuant to Chapter 1145 of the Statutes of 1990, after holding one or more public hearings. The updated model ordinance shall be based on the recommendations set forth in the report prepared pursuant to Chapter 682 of the Statutes of 2004 and shall meet the requirements of Section 65596.
(2) Before the adoption of the updated model ordinance pursuant to paragraph (1), the department shall prepare and submit to the Legislature a report relating to both of the following:
(A) The extent to which local agencies have complied with the model water efficient landscape ordinance adopted pursuant to Chapter 1145 of the Statutes of 1990.
(B) The department's recommendations regarding the landscape water budget component of the updated model ordinance described in subdivision (b) of Section 65596.
(b) Not later than January 31, 2009, the department shall distribute the updated model ordinance adopted pursuant to subdivision (a) to all local agencies and other interested parties.
(c) On or before January 1, 2010, a local agency shall adopt one of the following:
(1) A water efficient landscape ordinance that is, based on evidence in the record, at least as effective in conserving water as the updated model ordinance adopted by the department pursuant to subdivision (a).
(2) The updated model ordinance described in paragraph (1).
(d) If the local agency has not adopted, on or before January 1, 2010, a water efficient landscape ordinance pursuant to subdivision (c), the updated model ordinance adopted by the department pursuant to subdivision (a) shall apply within the jurisdiction of the local agency as of that date, shall be enforced by the local agency, and shall have the same force and effect as if adopted by the local agency.
(e) Nothing in this article shall be construed to require the local

agency's water efficient landscape ordinance to duplicate, or to conflict with, a water efficiency program or measure implemented by a public water system, as defined in Section 116275 of the Health and Safety Code, within the jurisdictional boundaries of the local agency.

65596. The updated model ordinance adopted pursuant to Section 65595 shall do all the following in order to reduce water use:

(a) Include provisions for water conservation and the appropriate use and groupings of plants that are well-adapted to particular sites and to particular climatic, soil, or topographic conditions. The model ordinance shall not prohibit or require specific plant species, but it may include conditions for the use of plant species or encourage water conserving plants. However, the model ordinance shall not include conditions that have the effect of prohibiting or requiring specific plant species.

(b) Include a landscape water budget component that establishes the maximum amount of water to be applied through the irrigation system, based on climate, landscape size, irrigation efficiency, and plant needs.

(c) Promote the benefits of consistent local ordinances in neighboring areas.

(d) Encourage the capture and retention of stormwater onsite to improve water use efficiency or water quality.

(e) Include provisions for the use of automatic irrigation systems and irrigation schedules based on climatic conditions, specific terrains and soil types, and other environmental conditions. The model ordinance shall include references to local, state, and federal laws and regulations regarding standards for water-conserving irrigation equipment. The model ordinance may include climate information for irrigation scheduling based on the California Irrigation Management Information System.

(f) Include provisions for onsite soil assessment and soil management plans that include grading and drainage to promote healthy plant growth and to prevent excessive erosion and runoff, and the use of mulches in shrub areas, garden beds, and landscaped areas where appropriate.

(g) Promote the use of recycled water consistent with Article 4 (commencing with Section 13520) of Chapter 7 of Division 7 of the Water Code.

(h) Seek to educate water users on the efficient use of water and the benefits of doing so.

(i) Address regional differences, including fire prevention needs.

(j) Exempt landscaping that is part of a registered historical site.

(k) Encourage the use of economic incentives to promote the efficient use of water.

(*l*) Include provisions for landscape maintenance practices that foster long-term landscape water conservation. Landscape maintenance practices may include, but are not limited to, performing routine irrigation system repair and adjustments, conducting water audits, and prescribing the amount of water applied per landscaped acre.

(m) Include provisions to minimize landscape irrigation overspray and runoff.

65597. Not later than January 31, 2010, each local agency shall notify the department as to whether the local agency is subject to the department's updated model ordinance adopted pursuant to Section 65595, and if not, shall submit to the department a copy of the water efficient landscape ordinance adopted by the local agency, and a copy of the local agency's findings and evidence in the record that its water efficient landscape ordinance is at least as effective in conserving water as the department's updated model ordinance. Not later than January 31, 2011, the department shall, to the extent funds are appropriated, prepare and submit a report to the Legislature summarizing the status of water efficient landscape ordinances adopted by local agencies.

65598. Any model ordinance adopted pursuant to this article shall exempt cemeteries from all provisions of the ordinance except those set forth in subdivisions (h), (k), and (*l*) of Section 65596. In adopting language specific to cemeteries, the department shall recognize the special landscape management needs of cemeteries.

65599. Any actions or proceedings to attach, review, set aside, void, or annul the act, decision, or findings of a local agency on the ground of noncompliance with this article shall be brought pursuant to Section 1085 of the Code of Civil Procedure. SEC. 4. Section 25401.9 is added to the Public Resources Code, to read:

25401.9. (a) To the extent that funds are available, the commission, in consultation with the Department of Water Resources, shall adopt by regulation, after holding one or more public hearings, performance standards and labeling requirements for landscape irrigation equipment, including, but not limited to, irrigation controllers, moisture sensors, emission devices, and valves, for the purpose of reducing the wasteful, uneconomic, inefficient, or unnecessary consumption of energy or water. (b) For the purposes of complying with subdivision (a), the commission shall do all of the following:

(1) Adopt performance standards and labeling requirements for landscape irrigation controllers and moisture sensors on or before January 1, 2010.
(2) Consider the Irrigation Association's Smart Water Application Technology Program testing protocols when adopting performance standards for landscape irrigation equipment, including, but not limited to, irrigation controllers, moisture sensors, emission devices, and valves.
(3) Prepare and submit a report to the Legislature, on or before January 1, 2010, that sets forth on a proposed schedule for adopting performance standards and labeling requirements for emission devices and valves.
(c) On and after January 1, 2012, an irrigation controller or moisture sensor for landscape irrigation uses may not be sold or installed in the state unless the controller or sensor meets the performance standards and labeling requirements established pursuant to this section.

SEC. 5. Article 4.5 (commencing with Section 535) is added to Chapter 8 of Division 1 of the Water Code, to read:

Article 4.5. Irrigated Landscape
535. (a) A water purveyor shall require as a condition of new retail water service on and after January 1, 2008, the installation of separate water meters to measure the volume of water used exclusively for landscape purposes. (b) Subdivision (a) does not apply to either of the following:
(1) Single-family residential connections.
(2) Connections used to supply water for the commercial production of agricultural crops or livestock.
(c) Subdivision (a) applies only to a service connection for which both of the following apply:
(1) The connection serves property with more than 5,000 square feet of irrigated landscape.
(2) The connection is supplied by a water purveyor that serves 15 or more service connections.
(d) For the purposes of this section, "new retail water service" means the installation of a new water meter where water service has not been previously provided, and does not include applications for new water service submitted before January 1, 2007.

SEC. 6. If the Commission on State Mandates determines that this act contains costs mandated by the state, reimbursement to local agencies and school districts for those costs shall be made pursuant to Part 7 (commencing with Section 17500) of Division 4 of Title 2 of the Government Code.

APPENDIX B

WEB-BASED WATER BUDGET CALCULATORS

- California Urban Water Conservation Council "Water Saver Home" site - http://www.h2ouse.net/
- The Irrigation Water Management Society - http://www.iwms.org/C_Water_Budget.html
- Green Industries of Colorado (GreenCO) - www.greenco.org/downloadables/GreenCO%20Calculator.xls
- Mission Springs Water District - http://www.mswd.org/forms/conservation_water_budget_calculator.pdf
- Irvine Ranch Water District - http://www.irwd.com/Conservation/allocation_calc.php?a=res
- Southern California Water Agencies - http://www.bewaterwise.com/calculator.html
- Government of Alberta, Canada - http://www.agric.gov.ab.ca/app19/calc/volume/waterallocation.jsp

APPENDIX C

CITY OF ALBUQUERQUE, NEW MEXICO WATER CONSERVATION LANDSCAPING AND WATER WASTE ORDINANCE

PART 1: WATER CONSERVATION LANDSCAPING AND WATER WASTE

§ 6-1-1-1 SHORT TITLE.

This article shall be known as the "Water Conservation Landscaping and Water Waste Ordinance."

(Ord. 18-1995)

§ 6-1-1-2 INTENT.

(A) To implement the outdoor water use recommendations of the Water Conservation Task Force, as called for in Resolution Bill No. R-58, Enactment No. 49-1992, adopted by the Council in May of 1992.

(B) To assist in reducing overall per capita water use in the city by 30%.

(C) To reduce yard irrigation and irrigation-related water waste, which comprise over 40% of the city's total annual water usage. To reduce peak summer usage, which is two to three times winter usage and determines the need for capital facilities to adequately meet system demand. To reduce irrigation water usage without sacrificing landscape quality by using lower water use plants, improved design and planting practices, different watering practices, and better irrigation system design and maintenance.

(D) To reduce water waste; i.e., overwatering, inefficient watering, or release of excess water which generates fugitive water in the public right-of-way. To reduce damage to publicly owned streets and the public expenditures necessary to repair the damage caused by this wasted water. To increase street safety by reducing the potential of frozen water on public right-of-way.

(E) To initially encourage voluntary water conservation for existing single-family residences while requiring conservation on all other properties. To apply more stringent requirements to city-owned facilities to set an example.

(Ord. 18-1995)

§ 6-1-1-3 DEFINITIONS.

For the purpose of this article, the following definitions shall apply unless the context clearly indicates or requires a different meaning.

ATHLETIC FIELD. A turf area used primarily for organized sports.

AUTOMATIC CONTROLLER. A solid state timer capable of operating valve stations to set the days and length of time water is applied.

BUBBLERS. Irrigation heads which deliver water to the soil adjacent to the heads.

CITY OWNED. Property owned by the City of Albuquerque.

COVENANTS. Agreements entered into by property owners, leaseholders, and renters which set conditions for the use, maintenance, and/or sale of property.

DEVELOPMENT. The construction, erection, or emplacement of one or more buildings, structures, or surface improvements on land which is a premises in order to establish or expand a principal residential or nonresidential use.

DISTURBED SLOPES. Slopes that have been altered from their natural configuration or vegetative cover by human activity.

DRIP IRRIGATION. Low pressure, low volume irrigation applied slowly, near or at ground level to minimize runoff and loss to evaporation.

EVAPOTRANSPIRATION. The quantity of water evaporated from adjacent soil surfaces and transpired by plants during a specific time.

EVEN-NUMBERED PROPERTIES. Properties whose official address ends in an even number, excluding city parks and golf courses. Landscaped areas associated with a building will use the number of that building as their address. Only one address shall be used for a large landscaped area associated with one building or activity, even if the landscaped area is broken into many separate subareas.

FLOW RESTRICTION DEVICE. Device applied by the water utility to the customer's meter that restricts the volume of flow to the customer.

FUGITIVE WATER. The pumping, flow, release, escape, or leakage of any water from any pipe, valve, faucet, connection, diversion, well, or any facility for the purposes of water supply, transport, storage, disposal, or delivery onto adjacent property or the public right-of-way.

HAND WATERING. The application of water for irrigation purposes through a hand-held hose, including hoses moved into position by hand and left to flow freely or through a shut-off nozzle.

HARVESTED WATER. Precipitation or irrigation runoff collected, stored and available for reuse for irrigation purposes.

HIGH WATER USE TURF. A surface layer of earth containing regularly mowed grass, with its roots, which requires large volumes and/or frequent application of water throughout its life. High water use grasses include but are not limited to varieties of Bluegrass, varieties of Ryegrass, varieties of Fescue, and Bentgrass.

INFILTRATION RATE. The amount of water absorbed by the soil per unit of time, usually expressed in inches per hour.

INSPECTION. An entry into and examination of premises for the purpose of ascertaining the existence or nonexistence of violations of this article.

LANDSCAPE AREA. The entire parcel less the building footprint, driveways, non-irrigated portions of parking lots and required off-street parking. Includes the public right-of-way.

LOW WATER USE PLANTS. Plants which are able to survive without supplemental water once established as specified in the "Albuquerque Plant List", published by the city.

MAYOR. The Mayor of Albuquerque or his/her designated representative.

MEDIUM AND LOW WATER USE TURF. A surface layer of earth containing regularly mowed grass, with its roots, which requires moderate or low volumes and/or frequency of application of water once established as specified in the "Albuquerque Plant List" published by the city. Low and medium water use grasses include but are not limited to Bermuda and Bermuda hybrids, Zoysia, blue grama, and Buffalo grass.

MEDIUM WATER USE PLANTS. Plants which require some supplemental watering throughout the life of the plant as specified in the "Albuquerque Plant List" published by the city.

MISTER. A device that produces a cooling effect by emitting fine particles of water into the air in the form of a mist.

MULCH. Any material such as leaves, bark, straw, or other materials applied to the soil surface to reduce evaporation.

NEW DEVELOPMENT. Any development approved by the Albuquerque Planning Department on or after October 1, 1995. For development for which landscaping is required, which is all development except single family residential, only that portion approved by the Albuquerque Planning Department on or after October 1, 1995 shall be considered new development. Development approved by the Albuquerque Planning Department prior to October 1, 1995, but not completed by October 1, 1998 shall also be considered new development.

NON-CITY OWNED. All property which is not owned by the City of Albuquerque.

ODD-NUMBERED PROPERTIES. Properties whose official address ends in an odd number, excluding city parks and golf courses. Large landscaped areas associated with a building will use the number of that building as their address. Only one address shall be used for a large landscaped area associated with one building or activity, even if the landscaped area is broken into many separate subareas.

PRECIPITATION RATE. The amount of water applied per unit of time, usually expressed in inches per hour.

PROPERTY HOLDER. An owner or leaseholder, whose landscaping is governed in whole or in part by rules applied to all property holders within a property holders' association.

PROPERTY HOLDERS ASSOCIATION. An association of property owners, leaseholders, or renters whose landscaping is governed in whole or in part by rules applied to all property holders within the development.

PUBLIC RIGHT-OF-WAY. The area of land acquired or obtained by the city, county, or state primarily for the use of the public for the movement of people, goods, vehicles, or storm water. For the purposes of this article the public right-of-way shall include curbs, streets, and storm water drainage inlets.

RESPONSIBLE PARTY. The owner, manager, supervisor, or person who receives the water bill, or person in charge of the property, facility, or operation during the period of time the violation(s) is observed.

RESTRICTED PLANTS. Plants which, as specified in the "Albuquerque Plant List" published by the city, are classified as restricted due to their high water use requirements and their potential for extensive use in landscaping. Restricted plants include high water use turf, clover, and Dichondra.

RUNOFF. Water which is not absorbed by the soil or landscape to which it is applied. Runoff occurs when water is applied too quickly (application rate exceeds infiltration rate), particularly if there is a severe slope. This article does not apply to stormwater runoff which is created by natural precipitation rather than human-caused or applied water use.

SHUT-OFF NOZZLE. Device attached to end of hose that completely shuts off the flow, even if left unattended.

SINGLE-FAMILY RESIDENTIAL. A lot or premises upon which is established one dwelling only. Of the allowable principal uses, such use shall be the only use on that lot or premises.

SPRAY IRRIGATION. The application of water to landscaping by means of a device that projects water through the air in the form of small particles or droplets.

SPRINKLER HEAD. A device that projects water through the air in the form of small particles or droplets.

STATIC WATER PRESSURE. The pipeline or municipal water supply pressure when water is not flowing.

TEMPORARY IRRIGATION SYSTEMS. Irrigation systems which are installed and permanently disabled within a period of 36 contiguous months.

VALVE. A device used to control the flow of water in the irrigation system.

WATER WASTE. The nonbeneficial use of water. Nonbeneficial uses include but are not restricted to:

(1) Landscape water applied in such a manner, rate and/or quantity that it overflows the landscaped area being watered and runs onto adjacent property or public right-of-way;

(2) Landscape water which leaves a sprinkler, sprinkler system, or other application device in such a manner or direction as to spray onto adjacent property or public right-of-way;

(3) Washing of vehicles, equipment, or hard surfaces such as parking lots, aprons, pads, driveways, or other surfaced areas when water is applied in sufficient quantity to flow from that surface onto adjacent property or the public right-of-way;

(4) Water applied in sufficient quantity to cause ponding on impervious surfaces on non-city owned property.

(Ord. 18-1995; Am. Ord. 24-1998; Am. Ord. 42-2001; Am. Ord. 13-2004)

§ 6-1-1-4 APPLICABILITY.

(A) Section 6-1-1-8, Water Budgets and Planting Restrictions, applies to all new development and to existing golf courses, city owned parks, and city owned athletic fields.

(B) Section 6-1-1-9, Design Regulations, applies to all new development and to major renovations of existing golf courses, city owned parks, and city owned athletic fields *originally constructed after 1971.*

(C) Section 6-1-1-10, Irrigation System Standards, applies to all new development and to expansions or major renovations of existing golf courses, city owned parks, and city owned athletic fields *originally constructed after 1971.* Single family residential shall be exempt from this section.

(D) Section 6-1-1-11, Inspection Requirements, applies to all new development.

(E) This article does not apply to water provided through the Middle Rio Grande Conservancy District for irrigation purposes. Water obtained through non-city water system sources, however, will be included in the calculation of inches per year for the water budgets for golf courses and parks, as described in Section 6-1-1-8.

(F) Certificates of occupancy for all new development except single family residential shall depend upon compliance with all requirements of this article.

(Ord. 18-1995; Am. Ord. 24-1998)

§ 6-1-1-5 WATERING RESTRICTIONS.

These restrictions apply to all properties within the city limits and/or served by the municipal water utility.

(A) All spray irrigation during the period beginning on April 1 and ending on October 1 of each year must occur between 6:00 p.m. and 10:00 a.m. beginning April 1, 2000. This restriction serves as a guideline for landscape watering on non-city owned property during 1999. This restriction shall not apply to drip irrigation and low precipitation bubblers, hand watering, or watering of containerized plants and plant stock.

(B) All spray irrigation on city owned property during the months of December through March must occur between 10:00 a.m. and 2:00 p.m. This restriction serves as a guideline for landscape watering on non-city owned property. This restriction shall not apply to drip irrigation and low precipitation bubblers, hand watering, or watering containerized plants and plant stock.

This restriction shall not apply to golf courses or parks that are in regular use or in use for a special event during these hours.

(C) Shutoff nozzles are required on any hoses used for hand watering, car washing or other outdoor uses, excepting hoses on single-family residential.

(D) All city owned properties other than parks and golf courses shall water no more than every other day. All even-numbered properties shall water only on even-numbered dates. All odd-numbered properties shall water only on odd-numbered dates. This restriction serves as a guideline for landscape watering on non-city owned property.

(E) Restrictions in divisions (A), (B) and (D) above do not apply to the following:

(1) Outdoor irrigation necessary for the establishment of newly sodded lawns and landscaping within the first 30 days of planting or watering of newly seeded turf within the first year of planting;

(2) Irrigation necessary for one day only where treatment with an application of chemicals requires immediate watering to preserve an existing landscape or to establish a new landscape;

(3) Water used to control dust or compact soil;

(4) Visually supervised operation of watering systems for short periods of time to check system condition and effectiveness.

(F) The city shall undertake an aggressive public information campaign to address the requirements of the spray irrigation restrictions for the remainder of 1999 and each year thereafter.

(G) 6-1-1-1 through 6-1-1-99 Water Conservation Landscaping and Water Waste shall be reviewed in its entirety in FY/04 as to its effectiveness and for necessary revisions. This evaluation will be incorporated into the FY/04 budget process.

(Ord. 18-1995; Am. Ord. 24-1998; Am. Ord. 54-1999) Penalty, see § 6-1-1-99

§ 6-1-1-6 WATER WASTE.

These restrictions apply to all properties within the city limits and/or served by the municipal water utility.

(A) No person, firm, corporation, or municipal or other government facility or operation shall waste, cause or permit to be wasted any water.

(B) No person, firm, corporation, or municipal or other government facility or operation shall cause or permit the flow of fugitive water onto adjacent property or public right-of-way.

(C) The restrictions in divisions (A) and (B) of this section do not apply to the following:

(1) Storm runoff allowed under provisions of the city's Drainage Ordinance as currently adopted or subsequently amended;

(2) Flow resulting from temporary water supply system failures or malfunctions. These failures or malfunctions shall be repaired as quickly as possible;

(3) Flow resulting from firefighting or routine inspection of fire hydrants or from fire training activities;

(4) Water applied as a dust control measure as may be required under Chapter 9, Article 5 of this code;

(5) Water applied to abate spills of flammable or otherwise hazardous materials, where water is the appropriate methodology;

(6) Water applied to prevent or abate health, safety, or accident hazards when alternate methods are not available;

(7) Flow resulting from routine inspection, operation, or maintenance of the municipal water supply system;

(8) Flow resulting from routine inspection or maintenance of irrigation systems;

(9) Water used by the Traffic Engineering Division, City of Albuquerque, in the course of installation or maintenance of traffic flow control devices;

(10) Water used for construction or maintenance activities where the application of water is the appropriate methodology and where no other practical alternative exists.

(Ord. 18-1995; Am. Ord. 24-1998) Penalty, see § 6-1-1-99

§ 6-1-1-7 SPECIAL PERMITS

These requirements apply to all properties within the city limits and/or served by the municipal water utility.

(A) Use of Misters

(1) The use of misters shall require a special permit, issued by the city. The Mayor shall develop regulations and administrative procedures for the issuance and conditions of such permits. The Mayor shall have the authority to limit the number of permits or revoke permits as deemed necessary to protect the public interest.

(2) Effective April 1, 1999, the use of misters without a permit, or in violation of permit conditions, shall constitute a violation of this article and shall be subject to the fee assessment processes described in §§ 6-1-1-13 and 6-1-1-99.

(3) Any person, firm, corporation, or municipal or other government facility selling, leasing, renting, installing or otherwise making misters available to any other person, firm, corporation, or municipal or other government facility shall provide notification to their customers of the special permit requirement for mister use. Notice may be delivered by prominently posting a sign at the point of purchase or by providing a document to each individual customer. The city shall provide approved language for such notification.

(Ord. 24-1998)

§ 6-1-1-8 WATER BUDGETS AND PLANTING RESTRICTIONS

Subsection (A) of this section applies to all city and non-city owned golf courses, and to all city owned parks and athletic fields. Subsection (B) of this section applies to all new development.

(A) Water Budgets for Parks and Golf Courses.

(1) Parks and golf courses shall use medium and low water use plants as much as possible. High water use turf or other restricted plants shall be allowed only in those areas with heavy usage or foot traffic, such as athletic fields, playgrounds, and golf course tees, greens, and fairways.

(2) All golf courses existing prior to October 1, 1995 will be allowed up to 40 inches of water per acre of landscape area per year. Golf courses using wells must report well usage to the city on a monthly basis. Any usage over the allowable amount will be subject to the excess use surcharge(s) described in division (A)(6) of this section. Usage will be calculated on a per individual golf course basis and shall include municipal and non-municipal water supplies.

(3) All new golf courses or existing golf course expansions permitted by the city after October 1, 1995 will be allowed up to 37 inches per acre of landscape area per year. Any usage over the allowable amount will be subject to the excess use surcharge(s) described in division (A)(6) of this section. Usage will be calculated on a per individual golf course basis and shall include municipal and non-municipal water supplies. The landscaped area for new golf courses shall not exceed 90 acres per 18 holes or 45 acres per 9 holes.

(4) All parks will be allowed up to 35 inches of water per acre of landscape area per year. Any usage over the allowable amount will be subject to the excess use surcharge(s) described in division (A)(6) of this section. Usage will be calculated on a per individual park basis and shall include municipal and non-municipal water supplies.

(5) Athletic fields will be allowed up to 45 inches per acre of landscape area per year. Any usage over the allowable amount will be subject to the excess use surcharge(s) described in division (A)(6) of this section. Usage will be calculated on a per individual athletic field basis and shall include municipal and non-municipal water supplies.

(6) Any usage over the approved water budget will be subject to the excess use surcharge(s) defined in the Water and Sewer Rate Ordinance as currently adopted or subsequently amended (see Ch. 6, Art. 4), and established by the Mayor's rules and regulations. This surcharge(s) will be calculated on an annual basis and applied to the February water bill for the property. If two different surcharges are defined in the Water and Sewer Rate Ordinance or the Mayor's rules and regulations, the surcharge for excess usage up to 10% of the water budget shall be the lower of the surcharges. The surcharge for excess usage over 10% of the water budget shall be the higher of the surcharges.

(7) For all parks, golf courses and other facilities with greater than ten acres of restricted plants, and developed after the effective date of this section, the owner or developer shall, when

available and economically feasible, use reclaimed wastewater, shallow groundwater or other alternative water supplies, as specified by the policies of the Albuquerque Water Resources Management Strategy.

(B) Planting Restrictions.

(1) All city owned new development other than parks, golf courses, and housing shall use medium and low water use plants on 100% of the landscape area.

(2) All city owned housing and all non-city owned properties other than golf courses shall not use high water use turf or other restricted plants on more than 20% of the landscape area, except that for single family residential properties;

(a) In the event that 20% of the landscape area is greater than 3,000 square feet, high water use turf and other restricted plants shall not be used on more than 3,000 square feet of the landscape area;

(b) In the event that 20% of the landscape area is less than 300 square feet, high water use turf and other restricted plants may be used on up to 300 square feet of the landscape area.

(C) Certain Restrictive Covenants Prohibited.

(1) A property holders' association shall not enforce a provision in a covenant that prohibits or restricts a property holder from:
(a) Removing turf grass and installing xeriscape landscaping in compliance with the restrictions for new development in subsection (B) of this section;
(b) Installing efficient irrigation systems, including underground drip systems; or

(c) Using rain barrels or other water harvesting devices, provided such devices adequately protect the public's health, safety, and welfare.

(2) A property holders' association may establish criteria for relandscaping to improve water use efficiency but cannot require a higher percentage of high water use turf than allowed in subsection (B) of this section except that it may require that the maximum percentage of high water use turf allowed in subsection (B) of this section be maintained.

(3) A property holders' association may establish criteria regarding type and placement of rainwater collection/harvesting.
(Ord. 18-1995; Am. Ord. 1-1998; Am. Ord. 24-1998; Am. Ord. 13-2004; Am. Ord. 41-2004; Am. Ord. 41-2004) Penalty, see § 6-1-1-99

§ 6-1-1-9 DESIGN REGULATIONS

The following regulations apply to all new development, and to expansions or major renovations as existing city owned parks, city and non-city owned golf courses, and city-owned athletic fields *originally constructed after 1971.*

(A) With the exception of temporary irrigation systems needed to establish low water use plants, spray irrigation shall not be used on slopes greater than four feet of horizontal distance per one foot vertical change (4:1).

(B) All existing disturbed slopes and all man-made slopes shall receive erosion control from plantings and/or terracing. Concrete, asphalt, or any other water and air impervious paving/cover will be allowed only where it is the most appropriate methodology and where no other practical alternative exists.

(C) Plants that require spray irrigation or a mowing frequency of more than three times per year shall not be used in street medians, except that spray irrigation may be used in street medians for up to 36 months where the primary objective is to reclaim disturbed areas with low water use plants.

(D) Spray irrigation shall not be used to apply water to any area within eight feet of a street curb or storm sewer inlet. These areas may be irrigated by drip, bubbler, soaker, or sub-surface irrigation systems.

(E) Sprinkler heads shall be installed at least eight inches away from impermeable surfaces.

(F) No spray irrigation shall be used in areas less than ten feet in any dimension excepting within back or side yards of residential properties, or where such an area is contiguous with adjacent property so that the dimension totals ten feet minimum. Within parking lots no spray irrigation shall be used on any area less than 15 feet in any dimension. These areas may be irrigated by drip, bubbler, soaker, or sub-surface irrigation systems.

(G) Any existing features should be evaluated for incorporation in design to include natural drainage, rock outcroppings, stands of native vegetation which can be protected, or detention areas where vegetation has grown and is being supported by nuisance flows or harvested water.

(H) The potential for using harvested water should be evaluated and, when practical, incorporated into landscape design. Such design shall be consistent with the requirements of the city's Flood Hazard Control Ordinance and the Drainage Ordinance as currently adopted or subsequently amended.

(I) Ponds, fountains, wetlands, marshes, water features for wildlife habitat, functional holding ponds or other reservoirs that are supplied in whole or in part by the municipal water supply shall not exceed 500 square feet or surface area unless approved by the Mayor. Multiple water features on the same property will be considered together to determine surface area. Flowing water used in fountains, waterfalls and similar features shall be recirculated.

(Ord. 18-1995; Am. Ord. 24-1998) Penalty, see § 6-1-1-99

§ 6-1-1-10 IRRIGATION SYSTEM STANDARDS

The following standards apply to all expansions or major renovations at existing parks, golf courses and athletic fields *originally constructed after 1971,* and to all new development except single family residential. The standards serve as voluntary guidelines for single-family residential development. In general, irrigation systems shall be designed to be site-specific, reflecting plant type, soil type, infiltration rates, slopes, and prevailing wind direction.

(A) Irrigation systems shall be designed to be in conformance with all provisions of this article. Temporary irrigation systems shall not be required to meet these standards.

(B) Application equipment for which the manufacturer specifies flow rates in gallons per minute (gpm) shall not share a control valve with equipment for which the manufacturer specifies flow rates in gallons per hour (gph). Irrigation systems shall be controlled by an automatic controller equipped with the following features:

(1) Two or more independent programming schedules;

(2) Capable of programming run times in one-minute increments and displaying the programmed run time as a numeric display;

(3) Total program memory retention;

(4) Ability to be fitted with an external rain switch interrupter and soil moisture sensor.

(C) No intentional overspray is allowed where it may obstruct pedestrian traffic on a city-required pedestrian walkway, as defined by the city's Sidewalk, Drive Pad, Curb and Gutter Ordinance as currently adopted or subsequently amended.

(D) Irrigation systems shall be designed such that water pressure at the sprinkler or emitter is not more than 20% in excess of the manufacturer's maximum recommended pressure range for that device. Pressure may be regulated by design or by the installation of a pressure regulating device or devices.

(E) Irrigation systems shall be designed to minimize low head line drainage.

(F) All new development with new spray irrigated landscaped areas totaling one-half acre or more shall have a Landscape Irrigation Audit performed by a Certified Landscape Irrigation Auditor, certified by the Irrigation Association. The auditor shall be independent of the property owner and of all contractors associated with the property. The audits will be conducted in accordance with the current edition of the Landscape Irrigation Auditor's Handbook. The minimum efficiency requirements to meet in the audit are a 60% distribution uniformity for all fixed spray systems and a 70% distribution uniformity for all rotary systems. The results of the audit shall be provided to the city in a letter or other form acceptable to the city and shall be signed by the Auditor. Compliance with this provision is required before the city will issue a Certificate of Occupancy or, in the case of park development, a Letter of Final Acceptance.

(G) All new development with spray irrigated landscapes greater than ten acres shall have the sprinkler heads tested for uniformity of performance using the Center for Irrigation Technology's (CIT) Sprinkler Profile and Coverage Evaluation (SPACE) program, or a comparable assessment acceptable to the city. The sprinkler heads shall have a scheduling coefficient of 1.3 or less for full circle heads and 1.5 or less for partial circle heads, with a rating of 1.0 being perfect. The sprinkler heads shall be installed in the spacing and pressure range tested. The results of this test shall be provided to the city in a form acceptable to the city.

Compliance with this provision is required before the city will issue a Certificate of Occupancy or, in the case of park development, a Letter of Final Acceptance.

(Ord. 18-1995; Am. Ord. 24-1998) Penalty, see § 6-1-1-99

§ 6-1-1-11 INSPECTION PROCEDURES

The following procedures apply to all new development:

(A) Inspection by Consent.

(1) Within the scope of his authority, the Mayor may conduct an inspection, with the voluntary consent of an occupant or custodian of the premises to be inspected who reasonably appears to be in control of the places to be inspected or otherwise authorized to give such consent.

(2) Before requesting consent for an inspection, the Mayor shall inform the person to whom the request is directed of the authority under and purposes for which the inspection is to be made and shall exhibit an identification card or document evidencing his authority to make such inspections.

(3) Inspections undertaken pursuant to this section shall be carried out with due regard for the convenience and privacy of the occupants, and during the daytime, unless, because of the nature of the premises, the convenience of the occupants, the nature of the possible violation or other circumstances, there is a reasonable basis for conducting the inspection at night.

(4) Notice of the purpose and approximate time of an inspection of an area not open to the general public shall be sent to the occupants or custodians of premises to be inspected not less than seven days before the inspection is undertaken.

(B) Inspection without Consent.

(1) Upon sufficient showing that consent to an inspection has been refused or is otherwise unobtainable with a reasonable period of time, the Mayor may make application for an inspection order/search warrant. Such application shall be made to a court having jurisdiction over the premises to be inspected. Such application shall set forth:

(a) The particular premises, or portion thereof sought to be inspected;

(b) That the owner or occupant of the premises has refused entry;

(c) That inspection of the premises is necessary to determine whether they comply with the requirements of this article;

(d) Any other reason necessitating the inspection, including knowledge or belief that a particular condition exists on the premises which constitutes a violation of this article; and

(e) That the Mayor is authorized by the city to make the inspection.

(2) The application shall be granted and the inspection order/search warrant issued upon a sufficient showing that inspection in the area in which the premises in question are located, or

inspection of the particular premises, is in accordance with reasonable legislative or administrative standards, and that the circumstances of the particular inspection for which application is made are otherwise reasonable. The court shall make and keep a record of the proceedings on the application, and enter thereon its finding in accordance with the requirements of this section.

(3) While executing the inspection order/search warrant the Mayor shall, if the premises in question are unoccupied at the time of execution, be authorized to use such force as is reasonably necessary to effect entry and make the inspection.

(4) While conducting the inspection the Mayor shall, if authorized by the court on proper showing, be accompanied by one or more law enforcement officers authorized to serve search warrants who shall assist the Mayor in executing the order at his direction.

(5) After execution of the order or after unsuccessful efforts to execute the order, as the case may be, the Mayor shall return the order to the court with a sworn report of the circumstances of execution or failure thereof.

(Ord. 18-1995; Am. Ord. 24-1998) Penalty, see § 6-1-1-99

§ 6-1-1-12 VARIANCES AND APPEALS

The Mayor shall be responsible for the enforcement of this article. The Mayor may prescribe policies, rules, or regulations to carry out the intent and purposes of this article.

(A) Variances to § 6-1-1-5 (Watering Restrictions) and § 6-1-1-6 (Water Waste), and § 6-1-1-7 (Special Permits).

(1) Administrative variances to the restrictions in §§ 6-1-1-5, 6-1-1-6, and 6-1-1-7 may be issued by the Mayor or his/her designee, only for the purposes of installing or retrofitting landscaping, provided that the general intent of this article has been met, compliance with this article is proven to cause practical difficulties and unnecessary hardship, and all options for abatement through modified water management have been exhausted. The criteria to determine hardship shall include level of capital outlay and time required to be in compliance with this article.

(2) Variances may be issued for a period not to exceed one year and shall stipulate both short-term corrective measures and a schedule for completion of long-term corrective measures. Variances issued to accommodate the installation or retrofitting of landscaping are only applicable to the site where the construction that will increase the possibility of wasted water is to occur. The variance shall apply only for the period of construction. As of the date of this legislation any existing variances shall be subject to these provisions. Variances must be renewed on an annual basis if long-term corrective measures cannot be completed within one year.

(B) Appeal of § 6-1-1-5 (Watering Restrictions), § 6-1-1-6 (Water Waste), and § 6-1-1-7 (Special Permits). Any responsible party may appeal fees for violations of §§ 6-1-1-5, 6-1-1-6, and 6-1-1-7 to the City Hearing Officer by filing an appeal within seven calendar days of receiving a notice of violation. Such request shall be made in writing and filed in the Office of the City Clerk. The appeal shall identify the property and state the grounds of appeal together with all material facts in support thereof. A filing fee of $20 shall be added to the water bill in the event the violation is upheld by the Hearing Officer. When a hearing is requested, the Hearing Officer shall send written notice by certified mail, return receipt requested, to the appellant of the time and place of the hearing. At the hearing the appellant shall have the right to present evidence as to the alleged fact upon which the Mayor based the determination of the need for assessment of fee or restriction of service and any other facts which may aid the Hearing Officer in determining whether this article has been violated. The Hearing Officer shall, within seven working days following the hearing, issue a written decision specifying the fee, if appropriate, and the action that must be taken to avoid additional penalty. Fees will be void and service will not be restricted if the written decision is not issued within seven working days.

(C) Judicial Review. The exclusive remedy for parties dissatisfied with the action of the City Hearing Officer on §§ 6-1-1-5, 6-1-1-6, and 6-1-1-7 shall be the filing of a petition for a writ of certiorari with the State District Court. The petition for review shall be limited to the record made at the administrative hearing held pursuant to this article.

(D) Variances to §§ 6-1-1-8 through 6-1-1-10 requirements. A variance to the regulations in §§ 6-1-1-8 through 6-1-1-10 may be issued by the Mayor, through the Zoning Hearing Examiner, provided that the general intent of this article has been met and compliance with this article is proven to cause practical difficulties and unnecessary hardship. The variance procedure for this article will comply with the variance procedure in the Zoning Code as currently adopted or subsequently amended. (This procedure is described in § 14-16-4-2.) Appeals of decisions of the Zoning Hearing Examiner are to the Environmental Planning Commission. Appeals of decisions of the Environmental Planning Commission are to the City Council. Appeal is made by filing written notice with the Planning Department within 15 days after the request for variance has been denied. Appeal procedures will comply with those in the Zoning Code, § 14-16-4-4.

(Ord. 18-1995; Am. Ord. 24-1998; Am. Ord. 49-2003)

§ 6-1-1-13 FEES; ASSESSMENT

(A) Fees and Restriction of Service. Any responsible party who violates any of the provisions of §§ 6-1-1-5, 6-1-1-6, and 6-1-1-7 shall be subject to progressively higher fees and flow restriction until the violation ceases or a variance is granted. The assessment of fees and application of flow restriction shall be consecutive for violations separated by less than three calendar years. Fees and flow restriction shall be suspended pending the outcome of an appeal or variance request.

(B) Assessment of Fees. Assessment of fees for violations of the regulations in §§ 6-1-1-5, 6-1-1-6, and 6-1-1-7 will be through the city utility bills for the responsible party's billing account. Fees shall be assessed to the account within 15 days following expiration of the appeal period or issuance of appeal findings and shall be listed as separate line item on the utility bill. Responsible parties shall be notified of the fee through certified mail within 15 days of the violation. Fees must be paid within the normal payment period allowed by the city utility billing system.

(C) In lieu of fees for violations of §§ 6-1-1-5 and 6-1-1-6, the responsible party may have a landscape water audit performed by an authorized landscape irrigation auditor, certified by the Irrigation Association. The audit will be conducted in accordance with the current edition of the Landscape Auditor's Handbook. The audit must be performed within 30 days of notification of violation and the audit recommendation must be implemented within 60 days of the audit. If these deadlines are not met, the fees for violation will apply.

(Ord. 18-1995; Am. Ord. 24-1998) Penalty, see § 6-1-1-99

§ 6-1-1-99 PENALTY.

(A) The schedule for assessment of fees and application of flow restriction for a violation of §§ 6-1-1-5, 6-1-1-6, and 6-1-1-7 shall be as follows:

(1) First observed violation – $20;

(2) Second observed violation – $50;

(3) Third observed violation - $100;

(4) Fourth observed violation - $300;

(5) Fifth observed violation - $400;

(6) Sixth observed violation - $600;

(7) Seventh observed violation - $800;

(8) Eighth observed violation - $1,000;

(9) Ninth or more observed violation: Either a $1,000 fee per violation plus application of a flow restriction device at meter or a $2,000 fee per each violation. The flow restriction device cannot be removed by the responsible party and will not be removed by the utility until the responsible party adequately demonstrates to the city that the violation has ceased or until a variance is granted.

(B) For the purpose of assessing fees or flow restriction for violations of §§ 6-1-1-5, 6-1-1-6, and 6-1-1-7, any previous violation shall not be considered if:

(1) A period of five years has elapsed since the violation was incurred; or

(2) The property is acquired by a new owner; or

(3) The violation occurred prior to July 1, 1998.

(C) Any responsible party who violates any provision of §§ 6-1-1-8 through 6-1-1-10 shall be deemed guilty of a misdemeanor, and upon conviction thereof, shall be punished by a fine not to exceed $500 and/or imprisonment for a period not to exceed 90 days. Application of fines for violations of the regulations in §§ 6-1-1-8 through 6-1-1-10 will comply with the Zoning Code as currently adopted or subsequently amended. (See §§ 14-16-4-1 through 14-16-4-12, and 14-16-4-99).

(D) Any person who violates the provisions of this article for which no other penalty is set forth, shall be subject to the general penalty provision of this code set forth in § 1-1-99.

(Ord. 18-1995; Am. Ord. 24-1998; Am. Ord. 42-2001; Am. Ord. 49-2003)

REFERENCES

Ann Arbor Water Utilities Department. 2003. Cost of Service Rate Study. Water Utilities Department. City Council Working Session. Ann Arbor, Mi. [Online] Available: <http://dallas.c-b.com/annarbor/council_draft.pdf>. [accessed January 30, 2006]

Aquacraft, Inc. (2005) Water and Energy Savings from High Efficiency Fixtures and Appliances in Single Family Homes. US EPA Draft Report.

AWWA (American Water Works Association). 2006. Water Conservation Programs – A Planning Manual. AWWA M52. Denver, Colorado.

AWWA (American Water Works Association). 2004. Avoiding Rate Shock: Making the Case for Water Rates. Denver, Colorado.

AWWA (American Water Works Association). 2005. Thinking Outside the Bill: A Utility Manager's Guide to Assisting Low-Income Water Customers. AWWA, Denver, Colorado.

AWWA (American Water Works Association). 2000. Manual of Water Supply Practices. Principles of Water Rates, Fees, and Charges. AWWA M1. Denver, Colorado

AWWA (American Water Works Association). 1991. Manual of Water Supply Practices. Water Rates. AWWA M1. Denver, Colorado

Bamzai, A., S. Silva, and T. Ash. 1997. The Outdoor Water Savings Potential of Price Incentives and Efficient Landscape Management. AWWA 1997 Annual Conference Proceedings. Atlanta, GA.

Barta, R. 2004. Stretching Urban Water Supplies in Colorado Strategies for Landscape Water Conservation. Colorado Water Resources Research Institute. Special Report 13. [Online]. Available: <http://cwrri.colostate.edu/pubs/series/specialreport/SR13-BARTA.pdf >. [accessed January 26, 2006]

Beecher, J.A., and P.C. Mann. 1991. Cost-Allocation and Rate Design for Water Utilities. Denver, Colorado: American Water Works Association Research Foundation.

Beecher, J.A., and P.C. Mann. 1990. Deregulation and Regulatory Alternatives for Water Utilities. Columbus, Ohio: The National Regulatory Research Institute.

Bennett, D. 1996. Water Conservation Savings and Drought Response. EBMUD Internal memo to Director of California Water and Natural Resources. Oakland, CA.

Berg, J. and T. Ash. 2002. Landscape Performance Certification Program. AWWA 2002 Water Sources Conference Proceedings. Las Vegas, NV.

Billings, R.B. and D.E. Agthe. 1980. Price Elasticities for Water: A Case of Increasing Block Rates. Land Economics, 56(1): 73-84.

Billings, R. B., and W. M. Day. 1989. Demand Management Factors in Residential Water Use: The Southern Arizona Experience. Journal AWWA, 81(3): 58-64.

Billingsley, W.G.. 2004. An Assessment of Municipal Drought Contingency Planning in Texas. 2004, AWWA Water Sources Conference Proceedings. (2004)

Bishop, D.B. and J.A. Weber. 1996. Impacts of Demand Reduction on Utilities. Denver, Colorado: American Water Works Association Research Foundation.

Bonbright, J.C., A.L. Danielson, and D.R. Kamerschen. 1988. Principles of Public Utility Rates. Arlington, Virginia: Public Utilities Report

Breyer, Stephen G,, R. B. Stewart, C. R. Sunstein, and M. L. Spitzer, *Administrative Law and Regulatory Policy: Problems, Text, and Cases,* Fifth Edition, Aspen Law & Business, 2002.

Brown and Caldwell Consulting Engineers. 1984. Residential Water Conservation Projects – Summary Report. Washington, D.C.: US Department of Housing and Urban Development (HUD).

Camp, R.C. 1978. The Inelastic Demand for Residential Water: New Findings. Journal AWWA, 70(8): 452-458.

Carney, D. et. al. 2003. Commercial Landscape Surveys – A New Standard of Excellence. AWWA 2003 Annual Conference Proceedings. Anaheim, CA.

Carver, P.H. and J.J. Boland. 1980. Short- and Long-Run Effects of Price on Municipal Water Use. Water Resources Research, 16(4): 609-616.

Cavanagh S.M., W.M. Hanemann, and R.N. Stavins. 2001. Muffled Price Signals: Household Water Demand Under Increasing-Block Prices. (Working Paper of December 2001.)

Centennial Water and Sanitation District. 2005. Rate Structure Q & A [Online]. Available: <http://www.highlandsranch.org/06_wsan/06_wsan_pdf/RatesFAQs.pdf>. [accessed January 25, 2006]

Chesnutt, T.W., J. Beecher, et al., Designing, Evaluating, and Implementing Conservation Rate Structures, A handbook for the California Urban Water Conservation Council, Sacramento, California, July 1997.

Chesnutt T. and J. Beecher. 1998. Conservation Rates in the Real World. Jour. AWWA, 90(2):67.

Chesnutt, T.W., C.N. McSpadden, and D.M. Pekelney. 1996. An Impact Evaluation of Home Water Surveys in Los Angeles, A report for the Metropolitan Water District of Southern California. Los Angeles, CA.

Chesnutt, T. and P. Mayer. 2004. Water Budget-Based Rate Structures: A New Look at an Old Idea. AWWA 2004 Water Sources Conference Proceedings. Austin, TX.

Chesnutt, T. and D. Pekelney. 2002. A Primer on Individualized Water Rates: Designing and Implementing Water Budget-based Rates. AWWA Water Sources 2002 Conference Proceedings.

City of Santa Rosa. 2001. Conservation Pricing and Santa Rosa's Water Rates [Online]. Available: <www.ci.santa-rosa.ca.us/wc/rate_study.asp> [accessed February 3, 2006]

Collinge, R. 1996. Conservation Feebates. Jour. AWWA, 88(1):70.

Costanza, F.B. 1983. "Considerations in Distributing Costs of Service to Customer Classes". Proceedings of AWWA Seminar on Water Rates: An Equitability Challenge. Denver, Colorado: American Water Works Association pp. 40-51.

Cummings, R., N. Norton, V. Norton, M.B. Walker, and D. Wilson. 2004. Conservation Pricing of Household Water Use in Georgia Communities: A Preliminary Exploration. Water Policy Working Paper 2004-011 [Online]. Available: <http://www.h2opolicycenter.org/pdf_documents/water_workingpapers/2004-011.pdf> [accessed January 30, 2006]

Cummings, R., M.B. Walker. 2006. Conservation Pricing Of Household Water Use In Rural Communities. Water Policy Working Paper 2006-001. Georgia Planning and Water Policy Center [Online]. Available: <http://www.h2opolicycenter.org/pdf_documents/water_workingpapers/WP2006-001.pdf> [accessed February 1, 2006]

Cuthbert, R.W. and P. R. Lemoine, 1996. "Conservation-Oriented Water Rates", AWWA Journal, 88(11):68-78

CUWCC. 2006. BMP 5: Large Landscape Conservation Programs and Incentives. California Urban Water Conservation Council. Sacramento, CA.

CUWCC. 2005. Water Smart Landscapes for California. Report to the Governor and the Legislature. California Urban Water Conservation Council. Sacramento, CA.

Danielson, L.E. 1979. An Analysis of Residential Demand For Water Using Micro Time-Series Data. Water Resources Research, 15(4): 763-767.

De La Piedra, J. 2004. Landscape and Agricultural Area Measurement and Water Use Budgets. AWWA 2004 Water Sources Conference Proceedings. Austin, TX.

Deister, A. D.. 2004. Drought Preparedness vs. Drought Response-Comparison of Costs and Benefits. AWWA Annual Conference Proceedings. (2004)

DeOreo, W.B.. 2006. The Role of Water Conservation in a Long-Range Drought Plan. Journal of the American Water Works Association. February 2006.

DeOreo, W.B., et. al. 2005. Why the City of Boulder Chose a Water Budget Rate Structure and Where They Go from Here. AWWA 2005 Annual Conference Proceedings. San Francisco, CA.

DeOreo, W.B., P.W. Mayer, and D.M. Lewis. 2000. Seattle Home Water Conservation Study. Boulder, CO: Aquacraft, Inc.

DeOreo, W.B., P.W. Mayer, E.L. Towler, and D.M. Lewis. 2003. Residential Indoor Water Conservation Study: Evaluation of High Efficiency Indoor Plumbing Fixture Retrofits in Single-Family Homes in the East Bay Municipal Utility District Service Area. Boulder, CO: Aquacraft, Inc.

DeOreo, W.B., P.W. Mayer, E.L Towler, and L.A. Martien. 2004. Tampa Water Department Residential Water Conservation Study. Boulder, CO: Aquacraft, Inc.

Duke, J. and R. Ehemann. 2002. The Conservation of Residential Water Scarcity Pricing of Water in Northern New Castle County. College of Agriculture and Natural Resources. University of Delaware. Newark, De. [Online]. Available: <http://www.udel.edu/FREC/PUBS/ER02-01x4.pdf> [accessed January 31, 2006]

Dziegielewski, B. et al. 2000. Commercial and Institutional End Uses of Water. . Denver, Colorado: American Water Works Association Research Foundation.

Espey, M., J. Espey, W.D. Shaw. 1997. Price Elasticity of Residential Demand for Water: a Meta-Analysis. Water Resources Research, 33(6): 1369-1374.

Florida Department of Environmental Protection. 2002. Water Conservation: Preventing and reducing wasteful, uneconomical, impractical, or unreasonable use of water resources. Florida Water Conservation Initiative. (Section 62- 40.412(1), F.A.C.). pages 55; 61-76. [Online]. Available: <http://www.dep.state.fl.us/water/waterpolicy/docs/WCI_2002_Final_Report.pdf>. [accessed January 27, 2006]

Gibbs, K.C. 1978. Price Variable in Residential Water Demand Models. Water Resources Research, 14(1): 15-18.

Goodman, J. 1999. Water Conservation from User Charges in Multi-family Rental Housing. Washington, D.C.: American Real Estate and Urban Economics Association.

Griffin, Ronald C. 2001. Effective Water Pricing. Journal of the American Resources Association. 37(5):1335-1347.

Hall, D. C. and W. M. Hanemann, 1996. Urban Water Rate Design based on Marginal Cost. *Advances in the Economics of Environmental Resources: Marginal Cost Rate Design and Wholesale Water Markets, Volume 1* JAI Press, Inc., Greenwich, Connecticut, pp. 95-122.

Hanemann, W.M. 1998. Determinants of Urban Water Use. In Urban Water Demand Management and Planning. Edited by D.D. Bauman, J.J. Boland, and W.M. Hanemann. New York: McGraw Hill, Inc.

Hewitt, Julie A. and W. Michael Hanemann. 1995. A Discrete/Continuous Choice Approach to Residential Water Demand under Block Rate Pricing. Land Economics, 72(2): 173-192.

Howe, C.W. 1982. The Impact of Price on Residential Water Demand: Some New Insights. Water Resources Research, 18(4): 713-716.

Howe, C.W. and F.P. Linaweaver. 1967. The Impact of Price on Residential Water Demand and Its Relation to System Design and Price Structure. Water Resources Research, 3(1): 13-32.

Hydrosphere Resource Consultants. 1999. City of Boulder Water Conservation Futures Study. City of Boulder, Colorado.

Ingram, P., M. Rosas, C. Chan. 2003. Water Metering and Rates. Urban Water Management Plan. Chapter 5. City of Redwood City. Public Works Services Department. [Online]. Available: <http://www.redwoodcity.org/publicworks/pdf/UWMP_June_2003_Chapter5.pdf>. [accessed January 30, 2006]

Integra Engineering. 2000. City of Boulder Treated Water Master Plan. City of Boulder, Colorado.

IRWD (Irvine Ranch Water District). 2005. Irvine Ranch Water District Schedule of Rates and Charges [Online]. Available: <http://www.irwd.com/AboutIRWD/budget_rates/ratescharges.pdf>. [accessed January 25, 2006]

Jensen et al. 1990 Evapotranspiration and Water Irrigation Water Requirements. American Society of Civil Engineers.

Jordan, J. L. Pricing to Encourage Conservation: Which Price? Which Rate Structure? University of Georgia. Available: <http://www.ucowr.siu.edu/updates/pdf/V114_A5.pdf>. [Accessed January 27, 2006]

Jordan, J.L. Issue 4: Issues in Water Pricing. Georgia Water Series. University of Georgia. Department of Agriculture and Applied Economics. Faculty Series 98-16. [Online]. Available: <http://www.griffin.uga.edu/water/fs98_16.pdf>. [accessed February 1, 2006]

Jordan, J. L. and R. Albani, 1998. An Evaluation of Conservation Rate Structures. Hagler Bailly, Arlington, Virginia.

Jordan, J. L. and R. Albani, 1999. Using Conservation Rate Structures, AWWA Journal, 91(8):66-73.

Kah, G. et. al. 1998. Landscape Area Measurement Supports Water Budgeting. The Irrigation Association, 1998 IA Technical Conference Proceedings. San Diego, CA.

LADWP (Los Angeles Department of Power and Water). 2004. Understanding the LADWP Bill. Water Rates [Online]. Available: http://www.ladwp.com/ladwp/cms/ladwp001155.jsp. [Accessed January 31, 2006]

LADWP. 1995. Los Angeles Water Rates. Ordinance No. 170435 [Online]. Available: <http://ladwp.com - ladwp001149.pdf>. [Accessed January 31, 2006]

Lampe, L.K., K. Hahn, and P. Kenel. 2003. Dealing with Drought: Real-Time Management of Supply and Demand. AWWA Annual Conference Proceedings, 2003.

Loaiciga, H.A. and S. Renehan. 1997. Municipal Water Use and Water Rates Driven by Severe Drought: A Case Study. Journal of the American Water Resources Association, December.

Mayer, P.W., E. Towler, W.B. DeOreo, E. Caldwell, T. Miller, et. al. 2004. National Multiple Family Submetering and Allocation Billing Program Study – Final Report. U.S. Environmental Protection Agency, Washington, D.C.

Mayer, P.W., W.B. DeOreo, E.M. Opitz, J.C. Kiefer, W.Y. Davis, B. Dziegielewski and J.O. Nelson. 1999. Residential End Uses of Water. Denver, Colorado: American Water Works Association Research Foundation.

Michelsen, A.M., J.T. McGuckin, and D.M. Stumpf. 1998. Effectiveness of Residential Water Conservation Price and Nonprice Programs. Denver, Colorado: American Water Works Association Research Foundation.

Morrisville Finance Department. 2006. Morrisville, N.C. Finance Department Utility Rates. Available: <http://www.ci.morrisville.nc.us/finance/utilityRates.asp> [accessed January 27, 2006]

NC League of Municipalities and the Environmental Finance Center at UNC-Chapel Hill. 2005. North Carolina Water and Sewer Rates and Rate Structures. Chapel Hill, NC. [Online]. Available: <http://www.efc.unc.edu/publications/pdfs/NC%20Water%20and%20Sewer%20Rates%20Summary.pdf>. [accessed February 2, 2006]

Nieswiadomy, M.L. 1992. Estimating Urban Residential Water Demand: Effects of Price Structure, Conservation, Education. Water Resources Research, 28(3): 609-15.

OWASA (Orange Water and Sewer Authority). Questions and answers about ... SEASONAL WATER CONSERVATION RATES. [Online] Available: http://www.owasa.org/pages/seasonalrates.htm [accessed February 4, 2006]

Pekelney, D. and T. Chesnutt. 1997. Landscape Water Conservation Programs: Evaluation of Water Budget Based Rate Structures. A report prepared for the Metropolitan Water District and the Municipal Water District of Orange County by A & N Technical Services, Inc.

Pint, E.M. 1999. Household Responses to Increased Water Rates During the California Drought. Land Economics, 75(2): 246-266.

Planning and Management Consultants. 1984. Influence of Price and Rate Structures on Municipal and Industrial Water Use. Fort Belvoir, VA: Institute for Water Resources, United States Army Corps of Engineers.

Raftelis, G.A. 1993. Water and Wastewater Finance and Pricing. Chelsea, Michigan: Lewis Publications, 2nd Edition.

Raftelis, G. 2002. Water and Wastewater Rate Survey. Raftelis Financial Consulting. Charlotte, NC.

Rao, A.R., et. al. 1995. Development of a Drought Warning System for Indiana. Proceedings of the 1995 First International Conference on Water Resources Engineering: Texas Water '95, Volume 1.

Reidy, K. 2005. From Drought Response to Water Conservation Ethic: Implementation of the Water Budget Concept in Aurora, Colorado. AWWA 2005 Annual Conference Proceedings. San Francisco, CA.

Rettie, M., G. Haupt, M. Isbell, B. McClung, and R. Grantham. 2005. Effective Practices to Select Acquire, and Implement a Utility CIS, a report prepared for the Awwa Research Foundation. 1P-5.75C-91071-07/05-CM.

Renwick, M., R. Green, C. McCorkle. 1998. Measuring the Price Responsiveness of Residential Water Demand in California's Urban Areas. A report prepared for the California Department of Urban Resources. Available: <http://rubicon.water.ca.gov/pedreport.pdf>. [accessed January 30, 2006]

Rubin, S. 2005. Census Data Shed Light on US Water and Wastewater Costs. Journal AWWA, Vol. 97 Issue 4, April 2005, Page(s) 99-110

Schlette, T. C., and D. C. Kemp. 1991. Setting rates to encourage water conservation. Water Engineering and Management, 138(5):25-29.

Schneider, M.L. and E.E. Whitlatch. 1991. User-Specific Water Demand Elasticities. Journal of Water Resources Planning and Management, 117(1): 52-73.

Shepard, A. 1998. Drought Contingency Planning: Evaluating the Effectiveness of Plans. Journal of Water Resources Planning and Management, ASCE. September/October.

Simjian, G. and F. Sanchez. 2004. Measuring Landscapes Through Aerial Photography: From Theory to Application. AWWA 2004 Water Sources Conference Proceedings. Austin, TX.

Stallworth, H. Water and Wastewater Pricing An Informational Overview. 2000. U.S. Environmental Protection Agency. Office of Wastewater Management. EPA 832-F-03-027.

Teodoro, Manuel. 2002. Tailored Rates. Jour. AWWA, 94(10):54.

Town of Cary. 2005. Water and Sewer Utility Rates. Cary, N.C. Available: <http://www.townofcary.org/depts/fndept/utilityrates06.pdf>. [accessed January 27, 2006]

Vickers, Amy. 2001. *Handbook of Water Use and Conservation*. WaterPlow Press, Amherst, Massachusetts.

Vista Consulting Group, Inc. 1996. Managing the Revenue and Cash Flow Effects of Conservation. Denver, Colorado: American Water Works Association Research Foundation.

Wade, William., et al. 1991. Cost of Industrial Water Shortages. Spectrum Economics. Chicago, IL.

Washington State Department of Health. 1995. Conservation-Oriented Rates for Public Water Systems in Washington. Report to Legislature. DOH Publication #331-133. [Online]. Available: <http://www.mrsc.org/Subjects/Environment/water/doh331-113.pdf>. [accessed February 1, 2006]

Wang, Young-Doo, W. J. Smith, Jr., J. Byrne. 2005. Water Conservation Oriented Rates: Strategies to Extend Supply, Promote Equity, and Meet Minimum Flow Levels. AWWA. ISBN 1-58321-327-9; Catalog Number 20562.

Western Resource Advocates and Utah Rivers Council. 2005. Water Rate Structures in Utah: How Utah Cities Compare Using This Important Water Use Efficiency Tool. [Online]. Available: <http://www.westernresourceadvocates.org/water/Utah%20Water%20Rate%20Analysis.pdf>. [accessed January 31, 2006]

Whitcomb, John. 2005. Florida Water Rates Evaluation of Single-Family Homes. Prepared for the Southwest Florida, St. John's River, South Florida, and Northwest Florida Water Management Districts.

Wilhite, D.A. 1997. State Actions to Mitigate Drought: Lessons Learned. Journal of the American Water Resources Association, October.

Wilhite, D.D.. 1997. Responding to Drought: Common Threads From the Past, Visions for the Future. Journal of the American Water Resources Association, October.

Wilhite, D.D., et al. 2005. Drought and Water Crises, Science, Technology, and Management Issues. Taylor and Francis Group.

ABBREVIATIONS

AB	Assembly Bill
AF	acre-feet
AGI	adjusted gross income
AMR	automatic meter reading
ASCE	American Society of Civil Engineers
AWC	average winter consumption
AWWA	American Water Works Association
AwwaRF	Awwa Research Foundation
BMP	best management practice
CCF	hundred cubic feet
CF	cubic feet
CI	commercial and institutional
CII	commercial, industrial, and institutional
CIS	customer information system
CUWCC	California Urban Water Conservation Council
DWP	Department of Water and Power
e.g.	for example
EBMUD	East Bay Municipal Utility District
EPA	Environmental Protection Agency
EP Act	Energy Policy Act
ET	evapotranspiration

ETo	Reference evapotranspiration rate
gal.	gallon
gcd	gallons per capita per day
GIS	geographical information system
gpd	gallons per day
gpf	gallons per flush
gpm	gallons per minute
gpsf	gallons per square foot
GWRP	Groundwater Recovery Project
HCF	hundred cubic feet
HOA	homeowner's association
i.e.	for example
Inc.	incorporated
IRWD	Irvine Ranch Water District
K_c	crop coefficient
K_d	density coefficient
K_l	landscape coefficient
K_m	microclimate coefficient
K_z	zone coefficient
kgal	thousand gallons
l.	liter
LA	landscape area

LADWP	Los Angeles Department of Water and Power
LF	Low-flow
lpf	liters per flush
min.	minute
MWD	Municipal Water District
NAICS	North American Industry Classification System
PAC	Project Advisory Committee
PC	personal computer
pop.	population
QA	quality assurance
QAQC	quality assurance and quality control
R^2	coefficient of determination
REUWS	Residential End Uses of Water Study
sf	square foot, single-family
SIC	Standard Industrial Classification system
SJC	San Juan Capistrano
St.	street
St. Dev.	Standard deviation
TW	Trace Wizard

ULF	Ultra-low-flush
ULFT	Ultra-low-flush toilet
UPS	United Parcel Service
US	United States
V_p	plant water requirement
WBBR	water budget-base rate
WD	Water District

IWA Publishing's authorised EU representative for General Product Safety Regulations is Diane D'Arras, 15 rue Duret, 75116 Paris, France, e-mail: safety@iwap.co.uk.

Printed and bound by CPI Group (UK) Ltd, Croydon, CR0 4YY
17/09/2026
02218804-0019